T5-CQF-497

SETON HALL UNIVERSITY

3 3073 10070580 3

CD14 in the Inflammatory Response

Chemical Immunology

Vol. 74

Series Editors   *Luciano Adorini*, Milan
Ken-ichi Arai, Tokyo
Claudia Berek, Berlin
J. Donald Capra, Oklahoma City, Okla.
Anne-Marie Schmitt-Verhulst, Marseille
Byron H. Waksman, New York, N.Y.

KARGER   Basel · Freiburg · Paris · London · New York ·
New Delhi · Bangkok · Singapore · Tokyo · Sydney

CD14 in the Inflammatory Response

SETON HALL UNIVERSITY
UNIVERSITY LIBRARIES
SO. ORANGE, NJ 07079-2671

Volume Editor   *Robert S. Jack*, Greifswald

23 figures, 2 in color, and 7 tables, 2000

 **KARGER**

Basel · Freiburg · Paris · London · New York ·
New Delhi · Bangkok · Singapore · Tokyo · Sydney

Chemical Immunology

Formerly published as 'Progress in Allergy'
Founded 1939 by Paul Kallos

● ●
Dr. Robert S. Jack

Institut für Immunologie und Transfusionsmedizin
Klinikum der Universität Greifswald, Greifswald, Germany

RC
583
.P7
v.74

Bibliographic Indices. This publication is listed in bibliographic services, including Current Contents® and Index Medicus.

Drug Dosage. The authors and the publisher have exerted every effort to ensure that drug selection and dosage set forth in this text are in accord with current recommendations and practice at the time of publication. However, in view of ongoing research, changes in government regulations, and the constant flow of information relating to drug therapy and drug reactions, the reader is urged to check the package insert for each drug for any change in indications and dosage and for added warnings and precautions. This is particularly important when the recommended agent is a new and/or infrequently employed drug.

All rights reserved. No part of this publication may be translated into other languages, reproduced or utilized in any form or by any means electronic or mechanical, including photocopying, recording, microcopying, or by any information storage and retrieval system, without permission in writing from the publisher.

© Copyright 2000 by S. Karger AG, P.O. Box, CH–4009 Basel (Switzerland)
www.karger.com
Printed in Switzerland on acid-free paper by Reinhardt Druck, Basel
ISBN 3–8055–6917–3

0

Contents

1 Introduction: Hunting Devils
R.S. Jack, Greifswald

5 Chemical Structure, Molecular Conformation, and Bioactivity of Endotoxins
U. Seydel, A.B. Schromm, R. Blunck, K. Brandenburg, Borstel

7 Chemical Structure of LPS
9 Relation of Chemical Structure of Endotoxin to Bioactivity
10 Physical Structure of LPS and Lipid A
11   Critical Micellar Concentration, Supramolecular Structure and Molecular Conformation
13   Phase States
14   Molecular Modelling, Conformational Studies, and Crystallographic Data
16 Interaction of Endotoxins with Target Cell Membranes
17 Relation of Physicochemical Parameters to Biological Activity
18 Possible Activation Mechanisms or How Can the Correlation between Molecular Conformation and Biological Activity Be Understood?
20 Acknowledgments
20 References

25 Structure/Function Relationships of CD14
F. Stelter, Greifswald

25 Molecular Characteristics of CD14
26 Functional Properties of CD14
28 Generation of Soluble CD14 and Soluble CD14 Isoforms
29 Identification of the LPS-Binding Site of Human CD14

33 CD14-Induced Cellular Signalling
35 Concluding Remarks
36 References

42 Lipopolysaccharide-Binding Protein
R.R. Schumann, E. Latz, Berlin

42 Lipopolysaccharide-Binding Protein and CD14: Their Potential Function within the
   Innate Immune System
44 LBP Is a Member of a Family of Structurally and Functionally Related Proteins
45 LBP Is Upregulated during the Acute Phase
47 Mechanisms of LBP Function
47   Recognition of Endotoxin Micelles
48   LBP Facilitates Monomerization of LPS and Promotes Cellular Activation
49   LBP Catalyzes LPS Transfer
50   LBP-Mediated LPS Detoxification Involves Lipoproteins
51   LBP and CD14 Act as Lipid Transfer Proteins
54 References

61 Role of CD14 in Cellular Recognition of Bacterial Lipopolysaccharides
R.L. Kitchens, Dallas, Tex.

62 Functional Consequences of LPS-CD14 Interactions
62   mCD14 Confers Sensitive Responses to LPS
64   sCD14 Can Augment or Inhibit Cell Responses to LPS
64 Modes of LPS Binding to CD14
64   CD14 Recognizes LBP-Opsonized Gram-Negative Bacteria
66   CD14 Recognizes Complexes Formed by LPS Aggregates and LBP
66   CD14 Recognizes LPS Monomers that Can Be Rapidly Formed by the
      Action of LBP
67   CD14 Structure-Function Relationships
68 Relationship of LPS Signaling to LPS Internalization and Intracellular Traffic
70 Lipid A Recognition/Response Elements beyond CD14
70   Insights from Inhibitory Mechanisms of LPS Partial Structures
71   Basis of Ligand-Specific Recognition of Lipid A Analogs
71   Candidates from the Toll-Like Receptor Family
72 Non-LPS Ligands of CD14
74 Insights into the Nature of Microbial Pattern Recognition Provided by CD14
76 Acknowledgments
77 References

83 Interactions of CD14 with Components of Gram-Positive Bacteria
R. Dziarski, Gary, Ind.; *A.J. Ulmer*, Borstel; *D. Gupta*, Gary, Ind.

83 Discovery of CD14-Dependent Cell Activation by Non-Lipopolysaccharide Stimulants
83 CD14 as the Receptor for Gram-Positive Cell Walls

85 CD14 as the Receptor for Peptidoglycan

85   Structure and Biologic Activity of Peptidoglycan

85   Function of CD14 as the Peptidoglycan Receptor

86   Binding of Peptidoglycan to CD14

88   Polymeric Peptidoglycan Is Required for Binding to CD14 and Cell Activation

88   The Regions of CD14 Involved in Binding to Peptidoglycan and Cell Activation

89 CD14 as the Receptor for Lipoteichoic Acids

90 CD14 as the Receptor for Mycobacterial Lipoarabinomannan

90 CD14 as the Receptor for Other Bacterial and Nonbacterial Polymers

90 Differential Activation of CD14-Negative Cells by Complexes of Soluble CD14 with Bacterial Cell Wall Components

93 Function of CD14 as a 'Pattern-Recognition Receptor'

94 Mechanism of Cell Activation by CD14

96 CD14-Independent Cell Activation by Gram-Positive Bacteria: Comparison with LPS

97 Signal Transduction Pathways Activated by Gram-Positive Bacterial Components: Comparison with LPS

100 Transcription Factors and Genes Activated by Gram-Positive Bacterial Components: Comparison with LPS

101 Acknowledgments

101 References

108 **Soluble CD14-Mediated Cellular Responses to Lipopolysaccharide**
R.I. Tapping, P.S. Tobias, La Jolla, Calif.

108 Activation of Endothelial Cells by LPS

110 Soluble CD14-Mediated Cellular Activation

111 Structural Requirements of sCD14-Mediated Cellular Activation

113 Mechanism of sCD14-Mediated Cellular Activation

116 Acknowledgments

116 References

122 **Non-Inflammatory/Anti-Inflammatory CD14 Responses: CD14 in Apoptosis**
C.D. Gregory, Nottingham

122 Apoptosis: Cell Death without Inflammation

125 The Apoptotic-Cell Surface

127 Macrophage Receptors Implicated in Apoptotic-Cell Clearance

129 Interactions between CD14 and Apoptotic Cells

130 Ligands

132 Signalling Mechanisms: The Apoptotic Cell as an Anti-Inflammatory Signalling Device

135 Conclusions

136 References

Contents   VII

141 **TNF in the Inflammatory Response**
D.N. Männel, B. Echtenacher, Regensburg

142 TNF Structure and Function
144 TNF as an Endotoxic Mediator
144   TNF in Experimental Sepsis
144   TNF in Human Sepsis
145   Anti-TNF Studies
146 TNF as an Antibacterial Host-Defense Mediator
146   From LPS to TNF
146   TNF Protects in Experimental Sepsis
147   TNF Producer Cells
148   Mast Cells as TNF Producers
149   Stimuli for Early TNF Release
151   Functions of Early TNF
152 TNF in Chronic Inflammatory Diseases
153 Conclusion
153 References

162 **Clinical Aspects: From Systemic Inflammation to 'Immunoparalysis'**
H.-D. Volk, P. Reinke, W.-D. Döcke, Berlin

163 The Beneficial Effects of Inflammation in Severe Infections
165 What Are the Mechanisms of Monocyte Deactivation?
169 Is IL-10 the Only Mediator of Monocyte Deactivation in ICU Patients?
170 What Is the Cellular Source of the High IL-6 Plasma Levels in Septic Patients with 'Immunoparalysis'?
170 How Can We Restore the Immune Responsiveness?
175 Conclusions
176 References

178 **Author Index**

179 **Subject Index**

Jack RS (ed): CD14 in the Inflammatory Response.
Chem Immunol. Basel, Karger, 2000, vol 74, pp 1–4

·······················

Introduction: Hunting Devils

The adaptive immune system has the capacity to generate a truly stupen-
dous number of different antigen-specific receptors and hence can, in principle,
provide defense against virtually any pathogen. However, there is a price to
be paid for this flexibility and sophistication and that is that it takes several
days to generate a protective response. Some means has to be made available
to hold an infection in check over this initial period. This is the responsibility
of the so-called innate immune system. Unlike the specific immune response
the innate system must make do with just a handful of receptors which, being
'hard wired', cannot be tailored to a particular pathogen. The system therefore
requires receptors directed to structures which are commonly found on microorganisms but which are not expressed on our own cells. One such structure
is lipopolysaccharide (LPS), the endotoxin of gram-negative bacteria. LPS is
a major component of the outer membrane of gram-negative bacteria and it
is essential for their survival. The innate immune system is able to detect
minute traces of LPS because it possesses a high-affinity LPS receptor called
CD14 which is expressed on the surface of monocytes and macrophages. The
interaction of LPS with cell-bound CD14 leads to the release of proinflammatory mediators. The importance of CD14 as a central component of the
innate system is underlined by the fact that a CD14-deficient mouse fails to
mount a normal inflammatory response to LPS.

Interaction of macrophage-bound CD14 with LPS results in activation
of the cell which releases proinflammatory mediators such as TNFα. These
mediators diffuse out into the neighboring tissue where they in turn activate
the endothelial cells lining the capillaries. Activated endothelial cells are in a
position to permit granulocytes to exit from the blood and these phagocytic
cells then head down chemokine gradients to the site of infection. There they
gobble up as many bacteria as they can lay their hands on and, by doing so,
hold the infection in check. This is a useful – indeed essential – first line of

defense but it has to be strictly controlled. An overactivation of the system may result in the release of excessive amounts of proinflammatory mediators. Localized, this may lead to chronic inflammation, but if systemic, a global activation of endothelial cells may ensue leading to leakiness of the capillaries. This may set the stage for a catastrophic drop in blood pressure, multi-organ dysfunction and other symptoms of septic shock.

So much for the briefest possible summary of this fascinating molecule which lies at the heart of the inflammatory response. However, as the Germans so rightly claim, the devil is to be found hiding amongst the details. Nowhere is this more true than in the CD14 story.

The first devil is to be found lurking among the LPS molecules released from gram-negative bacteria. As discussed by Seydel, Schromm, Blunck and Brandenburg, the LPS molecule has a peculiar structure which in turn results in peculiar physical chemical properties. Chief among these is that LPS forms micelles in aqueous solution. In these micelles that part of the LPS molecule which is recognized by CD14 is hidden within the micelle. The micelles of course are in equilibrium with a constant low concentration of monomers, but this concentration is so very low that there is a kinetic block to the activation of CD14-bearing cells. CD14 in other words is a high-affinity receptor for a molecule which, because of its physical chemistry, is essentially cryptic. We solve this problem for our monocytes and macrophages by providing a serum protein called lipopolysaccharide-binding protein (LBP) which is able to extract LPS from micelles and present it in an accessible form to CD14. The properties of this LBP protein are described in detail by Schumann and Latz.

The second devil hides among the mysteries of the CD14 structure. CD14 exists both as a membrane-bound (mCD14) and as a soluble molecule (sCD14) free in solution. Both forms are biologically active (Tapping and Tobias). Sadly neither an X-ray nor an NMR structure is as yet available for CD14 so we do not know precisely what the binding site really looks like. However, a considerable body of information is available principally from in vitro mutagenesis experiments which identifies the part of the molecule required for interaction with LPS (reviewed by Stelter and by Tapping and Tobias). This is a point of more than mere academic interest since the details of the structure of the binding site determine the range of possible ligands other than LPS which CD14 may bind. This leads us directly into the territory of the third devil.

He, or possibly she, can be faintly discerned in the fact that – as discussed by Dziarski and coworkers – CD14 can be clearly shown to interact in vitro not only with LPS but also with components of gram-positive bacterial cell walls. Since the innate immune system has very few receptors at its disposal this has led to the suggestion that CD14 may also be involved in defense against gram-positive organisms. To date, this remains a controversial point which will require appropriate in vivo experiments to resolve.

One of the great success stories of recent years is the unfolding tale of the exorcism of the fourth devil who is involved in the signalling problem. CD14 is bound to the cell surface via a GPI anchor and consequently has no intracellular domain with which an activation signal can be transmitted to the cell. Various means of overcoming this problem have been considered over the years, but none have been entirely satisfactory. Now it seems that at least one – perhaps not the only – means of achieving signalling through CD14 involves the mouse toll-like receptor 4 molecule (TLR4) which may be acting as a coreceptor. The current status of this rapidly changing signalling story is laid out by Kitchens.

Though champagne and cigars are certainly merited for the unmasking of TLR-4, a word of caution is appropriate. This is because of the important recent discovery that CD14 is involved in the clearance of apoptotic cell remains by macrophages (reviewed by Gregory). There are really two quite separate surprises here. The first is that CD14 is involved in a biological process so remote from the inflammatory response. The second is that this CD14-dependent interaction far from resulting in an inflammatory response has the exact opposite effect. The response involves the release of mediators which actively inhibit inflammation. It rather looks as if CD14 may have the capacity to use more than one signal transduction route and that the choice of route used will be influenced by the ligand. TLR-4 is clearly important, but it may not be the whole story. Other possible signal transduction routes are discussed by Seydel by Kitchens. I have little doubt that the next couple of years will show that one or more devils is almost certainly still skulking in these CD14 signalling thickets.

In attempting to understand the workings of an inflammatory response we have to bear in mind that the fifth devil devotes himself to concentration-dependent events. Thus, LBP can act as an activator of CD14-positive cells when present at low concentration and – presumably by sequestering LPS – can inhibit cellular activation both in vitro and in vivo when present at acute phase levels. In other words, LBP may be proinflammatory under one set of circumstances and anti-inflammatory under others (Schumann and Latz). This remarkable concentration dependence of function is also seen with the soluble CD14 molecule. As reviewed by Tapping and Tobias, low concentrations of sCD14 can, under appropriate conditions, lead to activation of cells which do not themselves express membrane-bound CD14. In contrast, massive concentrations of sCD14 can block cellular activation presumably by sequestering LPS. Any attempt to manipulate the inflammatory system must take account of the ambivalent nature of these central components.

The sixth devil hides in the complexity of the system. Much of what we now know about CD14 function has been worked out in in vitro experimental systems. However, in a system as complex as the inflammatory response it is essential to review in vivo the conclusions drawn from in vitro experiments. Numerous cases of this are to be found in the reviews in this volume. One excellent example involves the long-standing idea – emanating from in vitro experiments – that in situations of acute or chronic inflammation TNFα may be regarded as merely deleterious. As discussed by Männel and Echtenacher, in vivo experiments demonstrate that this viewpoint is wrong. In a similar way in vitro experiments have clearly demonstrated that LBP can act as a transfer protein for various phospholipids leading to the idea that it may play a role in generalized lipid metabolism. However, analysis of an LBP knock-out mouse provides no evidence for any nonredundant role of LBP in lipid metabolism in vivo. By the same token, the CD14 ko mouse should provide an excellent experimental tool with which to probe the role of CD14 in apoptosis in vivo.

As outlined in these chapters, cellular activation via CD14 leads in most – but not all – situations to a release of proinflammatory mediators. A full discussion of these is well beyond the scope of this volume but the most important and best understood proinflammatory cytokine – TNFα – is discussed by Männel and Echtenacher. The TNFα story is of particular interest because it shows how closely basic research in this area impinges on potential clinical applications. This idea is extended in the final chapter in which Volk and coworkers describe the advances made in recent years in understanding the events taking place in septic shock – the most serious, and till now intractable form of acute inflammation known. Here, a combination of basic research plus the analysis of the disappointing clinical trials of anti-inflammatory strategies have led to the development of new clinical concepts which hold great promise for the future.

Robert S. Jack, Greifswald

Jack RS (ed): CD14 in the Inflammatory Response.
Chem Immunol. Basel, Karger, 2000, vol 74, pp 5–24

....

Chemical Structure, Molecular Conformation, and Bioactivity of Endotoxins

Ulrich Seydel, Andra B. Schromm, Rikard Blunck, Klaus Brandenburg

Research Center Borstel, Center for Medicine and Biosciences, Borstel, Germany

Gram-negative bacteria express at their surface a glycolipid called lipo-polysaccharide (LPS). LPS consists of a polysaccharide portion – subdivided into the O-specific chain (O-antigen) and the core oligosaccharide – which is covalently bound to the lipid portion termed lipid A which anchors the molecule in the outer leaflet of the outer membrane [1]. LPS participate in the physiological membrane functions and are, therefore, essential for bacterial growth and viability [2]. They represent a primary target for the interaction with antibacterial drugs and with components of the immune system of the host and have been shown to represent a high permeability barrier in particular for large hydrophobic agents [3]. Furthermore, when released from the bacterial surface, LPS play an important role in the pathogenesis and manifestation of gram-negative infection, in general, and of septic shock, in particular, and are thus called endotoxins [4]. On the other hand, they are also capable of producing beneficial effects in higher organisms depending upon amount and route of introduction [5]. Since the lipid A moiety is responsible for the major part of the biological activities, it is called the 'endotoxic principle' of LPS.

The biological activity of endotoxins may be discussed in the context of the interaction of these molecules with the membranes of host cells. The direct interaction of endotoxin molecules with host cell membranes via protein binding and/or hydrophobic interaction is considered to be an important step in the initiation of biological effects [6, 7] such as cytokine (TNFα and the interleukin family) production by monocytes/macrophages. However, not all LPS or lipid A molecules are equally active in this respect. Lipid A from

Escherichia coli carrying six fatty acids in asymmetric distribution was found to be most active. Reduction of biological activity down to complete inactivity was observed for LPS from various nonenterobacterial sources with lipid A structures deviating from that of *E. coli* in one or more of the following properties: (i) number and distribution of acyl chains linked to the sugar backbone; (ii) number and location of charges within the backbone region, and (iii) number of hydroxyl groups in the hydrophobic moiety [8]. Moreover, some of these lipid A variants exhibited strong antagonistic activity, i.e. they inhibited the agonistic action of biologically active LPS [9, 10].

Endotoxins as amphiphilic molecules form supramolecular aggregates in aqueous environments above a critical concentration (critical micellar concentration, CMC), which depends on their hydrophobicity, i.e. on the particular primary chemical structure [11]. The structure of these polymers can be micellar, lamellar or nonlamellar inverted, e.g. cubic or hexagonal II, depending on the conformation of the contributing molecules, which is again determined by the chemical structure. Polymer structure is influenced by ambient conditions such as temperature, pH, water content, and cation concentrations. For endotoxins, only lamellar or nonlamellar inverted structures have so far been found (overview see [12]). From the determination of the aggregate structure, the conformation of the isolated molecules can be deduced. It is cylindrical in the case of those lipid A, which form lamellar structures (identical cross-section of the hydrophilic and hydrophobic region), and conical for those lipid A, which adopt nonlamellar inverted structures (cross-section of the hydrophobic moiety larger than that of the hydrophilic one). Furthermore, the molecular conformation of a given LPS or lipid A molecule within an aggregate depends, among other things, on the fluidity (inversely correlated to the state of order) of its acyl chains, which can assume two phase states, the gel (β) phase and the liquid-crystalline (α) phase. Between these two phase states, a reversible transition takes place at a given phase transition temperature T_c, which depends, in the first place, on the length and degree of saturation of the acyl chains and on the conformation of the headgroup region and the distribution of charge density within it [11, 12].

Already in the 1970s, endotoxicologists became aware of a possible correlation between biological activity and aggregate structure of LPS [13]. However, at that time the chemical structures and possible contaminations were largely unknown. Later on, other groups continued in this direction applying more sophisticated methods [14–27]. Today, we think that the determination of the molecular conformation of endotoxin molecules is a most promising way to understand the molecular mechanisms underlying endotoxin bioactivity. There is a wealth of evidence that at least in certain parts of the activation pathway, endotoxin monomers or small oligomers are the biologically active units:

Firstly, the observed direct hydrophobic interaction of endotoxins with target cell membranes [6, 28, 29] should proceed efficiently only via the intercalation of monomers [11]. Secondly, the lipopolysaccharide-binding protein (LBP), which enhances biological activity of LPS [30], is known to disaggregate the endotoxins and to transport them directly into the host cell membrane [31]. Thirdly, in another pathway LBP transports endotoxin to the membrane-bound receptor protein, mCD14 [32]. There is experimental evidence that only a low number of endotoxin molecules is bound by each CD14 molecule [33]. CD14 is GPI-anchored and, thus, lacks a transmembrane domain. It can therefore not be directly involved in transmembrane signaling, but may be considered as an enhancer of biological action by binding monomerized endotoxin molecules and subsequently inserting them into the membrane. It has been demonstrated that, in addition to the molecular conformation, the presence of negative charges within the lipid A backbone is an important prerequisite for recognition and transport of endotoxins by LBP [31, 34]. The subsequent steps leading to cell activation are still unknown. It may be hypothesized that an interaction of endotoxin molecules with a transmembrane protein via H-bonds and/or electrostatic forces between the anionic lipid A portion and cationic protein takes place. Cellular signaling is then triggered by those LPS, which have a lipid A with a conical conformation by exerting steric stress on the signaling protein [31].

Chemical Structure of LPS

In the case of *Enterobacteriaceae*, LPS consists of three structural units, the O-specific chain, the core region, and the lipid A component (fig. 1). The O-specific chain is unique and characteristic for a given LPS serotype and, therefore, highly variable among LPS molecules of different serotypic origin [4, 35]. The core portion is structurally less variable and may be divided into the O-chain-proximal outer core and the lipid A-proximal inner core. Common elements present in the outer core are the pyranosidic hexoses *D*-glucose, *D*-galactose, 2-amino-2-deoxy-*D*-glucose (GlcN), and 2-amino-2-deoxy-*D*-galactose (GalN). The inner core is composed of 3-deoxy-*D*-manno-oct-2-ulosonic acid (dOclA, also termed 2-keto-3-deoxyoctulosonic acid, Kdo) and heptose residues, in general of the *L*-glycero-*D*-manno configuration, the latter often being phosphorylated. LPS expressing the complete core and the O-specific chain are termed wild-type or smooth (S-form) LPS, those lacking the O-chain are referred to as rough mutant (R-) LPS. LPS carrying only the Kdo units bound to lipid A are classified as deep rough mutant (Re-) LPS (see fig. 1 for *Salmonella minnesota*).

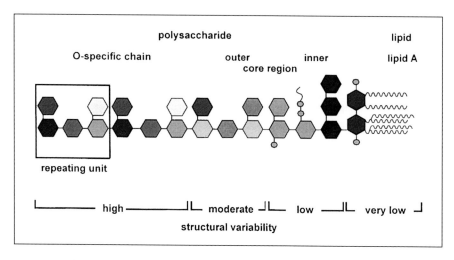

Fig. 1. Schematic chemical structure of lipopolysaccharide from *S. minnesota.*

Chemically, lipid A of many gram-negative bacteria, in particular *Entero-bacteriaceae*, consists of a bisphosphorylated β1,6-linked *D*-GlcN disaccharide, which carries up to seven acyl groups. This structure is highly conserved among bioactive LPS [4]. As an example, figure 2A shows the chemical structure of the lipid A component of *E. coli*. Six fatty acids are nonsymmetrically distributed over the two GlcN residues. An ester-linked (*R*)-3-hydroxymyristic acid at C3′ carrying myristic acid at its 3-OH group and an amide-linked (*R*)-3-hydroxymyristic acid, which is acylated by lauric acid at its 3-OH group are attached to the nonreducing GlcN residue (GlcN II). The reducing GlcN residue (GlcN I) carries 2 mol of (*R*)-3-hydroxymyristic acid, one in ester (position 3) and one in amide (position 2) linkage. Previously, it was assumed that the lipid A structure of most LPS from gram-negative bacteria should be constant ('conservative structure'). However, it is now known that nonenterobacterial LPS have – as a rule – a rather different lipid A structure. The dissimilarities are due to variations in the number, location, and chain length of the fatty acids, the substitution of phosphate groups, and the presence of 2,3-diamino-2,3-dideoxy-*D*-glucose (GlcN3N) instead of GlcN [8]. For example, lipid A from *Chromobacterium violaceum* carries a symmetrical fatty acid distribution with – on the average – shorter acyl chains [4] (fig. 2B). Also, there is phosphate-free lipid A, like that from *Rhodospirillum fulvum* [36], which carries a galacturonic acid instead of the 1-phosphate and a heptose instead of the 4′-phosphate, and have on average longer acyl chains (fig. 2C). These differences in the chemical structures considerably influence the physical

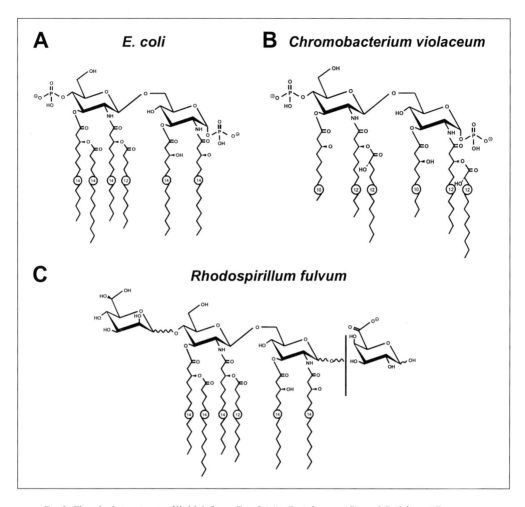

Fig. 2. Chemical structures of lipid A from *E. coli* (*A*), *C. violaceum* (*B*), and *R. fulvum* (*C*).

aggregate structures of these LPS types as well as their biological activities. This will be discussed in the next paragraphs.

Relation of Chemical Structure of Endotoxin to Bioactivity

Rough and smooth forms of LPS from various bacteria as well as different lipid A and lipid A partial structures have been investigated to characterize the bioactive regions of LPS in more detail. An important experimental basis for these investigations was the successful chemical synthesis of lipid A and corresponding lipid A partial structures, such as *E. coli*-type lipid A (compound 506) or precursor Ia (compound 406, or lipid IVa) carrying only four (*R*)-3-hydroxymyristic acid residues in symmetric distribution [37]. Full biological activity is expressed by a lipid A molecule consisting of a hexaacylated, bisphosphorylated β1,6-linked *D*-GlcN disaccharide, i.e. an *E. coli*-type lipid A [38]. Lipid A partial structures deficient in one of these elements are less active or even inactive in the induction of cytokines in human monocytes. For instance, the synthetic 1-dephospho and the 4′-dephospho lipid A partial structures were less active than compound 506, highlighting the significance of the phosphoryl groups (or the charges per se) for the biological activity of lipid A. In a systematic investigation, we have measured the IL-6 induction of peripheral blood mononuclear cells (PBMC) induced by different Re-LPS and lipid A partial structures with the main variation concerning the number of charges [34]. We found strong variations in the biological activity with the lowest activity for the compounds lacking phosphate such as synthetic compound '503' and lipid A from *R. fulvum*. Also, dephosphorylated Re-LPS which still has negative charges (two carboxylates in the Kdo region) was essentially inactive. Furthermore, it was found that the lipid A precursor Ia (compound 406) is completely inactive with respect to the induction of cytokine release in human monocytes [39]. It appears that the location of the secondary acyl residues is of importance as shown by the low bioactivity of a synthetic compound with a symmetrical distribution of the fatty acids. This was found to be similarly true for natural lipid A from *C. violaceum* [unpubl. results].

A variety of lipid A, lipid A analogues and nonbioactive synthetic partial structures has been tested with respect to LPS-induced reactions in vitro and in vivo. We and others have shown that among these the lipid A precursor Ia (compound 406) is a very effective LPS antagonist inhibiting LPS-induced cytokine production [10, 40]. Similar antagonistic effects were also described for the pentaacyl bisphosphoryl lipid A isolated from *Rhodobacter sphaeroides* [41], *Rhodobacter capsulatus* [9], and *C. violaceum* [own results, submitted].

Physical Structure of LPS and Lipid A

The most important physicochemical parameters, which seem to be relevant for an understanding of the influence of the above-mentioned variations in chemical structure on endotoxin bioactivity, are the critical micellar concentration (CMC), the phase states (fluidity of the endotoxin acyl chains), which are properties of the bulk lipid, and the molecular conformation, which is a property of the individual molecules. The molecular conformation can be deduced from the type of supramolecular aggregate structure built up from a large number of identical endotoxin molecules. Lipid A, as the endotoxic principle of LPS, constitutes the molecular entity which primarily interacts with the host cell membrane. Therefore, the determination of the conformation of the lipid A moiety of LPS seems to be the prerequisite for an understanding of biological action, whereas the role of the sugar moiety of LPS should be restricted to a modulation of lipid A bioactivity (the sugar has an influence on the hydrophobicity of endotoxin molecules and, with that, on their CMC as well as on the fluidity of the lipid A acyl chains). Thus, a more detailed physicochemical characterization of these features both of lipid A and also of LPS is an important issue in the characterization of the parameters which govern endotoxicity and determine why some structures are biologically highly active and others not (for extensive review, see also [12]).

Critical Micellar Concentration, Supramolecular Structure and Molecular Conformation

Unfortunately, no exact value for the CMC of lipid A has so far been determined due to the extreme experimental difficulties of working in the low concentration range, e.g. detection limits, loss of material due to absorption to walls. A rough estimation can be given from the limited data available for other lipids. Thus, given the change in CMC from $5\cdot10^{-10}$ M for dipalmitoyl-phosphatidylcholine to $7\cdot10^{-6}$ M for lysopalmitoylphosphatidylcholine [11] (which have the same headgroup but differ in the number of acyl chains by a factor of two), a value of well below 10^{-10} M may be assumed for hexaacyl lipid A since a CMC of $<10^{-7}$ M has been determined for the tetraacyl lipid A precursor IVa [42]. This approximation is supported by data on the solubility of Re-LPS determined experimentally applying equilibrium dialysis by Takayama's group [43, 44]. The authors found saturation values of the solubility of $3.3\cdot10^{-8}$ M at 22 °C and $2.8\cdot10^{-8}$ M at 37 °C. As the monomer concentration decreases with increasing aggregate concentration [11], these measurements would imply even higher values for the CMC than the values listed above. As Re-LPS should be less hydrophobic than lipid A, the CMC of the latter should be lower.

Above the CMC, the type of supramolecular structure found is governed by the molecular conformation (fig. 3). Long-range order within the aggregate structures has been studied using X-ray or neutron small-angle diffraction. Emmerling et al. [14], Wawra et al. [15], Labischinski et al. [16], and Naumann et al. [45] found similar results with dried or hydrated (max. 30% water content) samples of wild-type LPS, Re-LPS from *S. minnesota* R595, LPS K12, and lipid A. All authors found only one ordered lamellar phase in the temperature

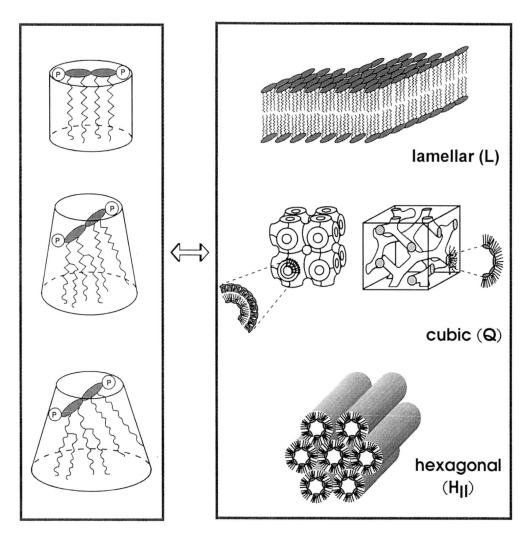

Fig. 3. Relation between molecular conformations (left) and supramolecular aggregate structures (right).

range 0–50 °C. Similarly, experiments with dry foils of synthetic lipid A analogs and partial structures – differing in the number of the hydrocarbon chains attached to an identical bisphosphoryl diglucosamine backbone – led, in each case, to diffraction patterns typical for lamellar structures [17]. In experiments at a water content >70%, however, these and other synthetic compounds analyzed with infrared spectroscopy and, for calibration, in some experiments also with X-ray diffraction [46] gave rather different results. Di- and triacyl lipid A analogs adopted micellar, tetra- to pentaacyl samples lamellar, and the others cubic or H_{II} structures. In neutron diffraction experiments using monophosphoryl lipid A from Re-LPS of *S. minnesota*, Labischinski et al. [18] found that at a water content of 99% again only lamellar structures were formed. In contrast, Hayter et al. [19] could best fit their data for Re-LPS from *E. coli* D21f2 to a randomly coiled tubular micelle structure.

The disparate results presented so far reflect the different experimental conditions used. In particular the strong lyotropism of endotoxins has great influence here. Thus, at 75% water content one Re-LPS molecule binds approximately 400 water molecules, one Ra-LPS binds 730 [21], whereas the well-hydrating phospholipid lecithin binds around 50 water molecules. This implies that the type of aggregate structure formed as well as the acyl chain order and the value of T_c depend on the water content.

Some of the data obtained so far under near physiological conditions can be summarized as follows: In pure lipid/water systems and at physiological water concentrations, free lipid A of the LPS of *E. coli* and *S. minnesota* aggregate into nonlamellar cubic structures already below T_c. With the beginning of the acyl chain melting process, free lipid A assumes nonlamellar cubic structures. With the completion of chain melting, the cubic structures change into the inverted hexagonal H_{II} form [25, 47].

In further studies, measurements of the aggregate structure were extended to other lipid A samples, including enterobacterial LPS and lipid A in different salt forms, monophosphoryl lipid A, and lipid A of nonenterobacterial sources like those of *R. capsulatus, Rhodopseudomonas viridis, Rhodocyclus gelatinosus, R. fulvum, Campylobacter jejuni* and *C. violaceum* [34, 48–50]. The measurements were performed exclusively under near physiological conditions with the purpose of directly correlating the results to data from biological analyses. It was found that the type of counterions (endotoxins in different salt forms) influenced the aggregate structure, and that the different nonenterobacterial lipid A samples showed a variety of aggregate structures: Lipid A from *R. gelatinosus* adopted H_{II} structures, that of *C. jejuni* and monophosphoryl lipid A of enterobacterial strains mixed lamellar/cubic, and those of *C. violaceum, R. capsulatus, R. viridis,* and *R. fulvum* pure lamellar structures. As stated above, these different aggregate structures correspond to different conformations of the isolated molecules. It was shown subsequently, that these different conformations are important for an understanding of the expression of biological activity as outlined later.

Phase States
For amphiphilic compounds like lipid A and LPS, an endothermic transition between a highly ordered gel phase (β phase) and a less-ordered liquid-

crystalline phase (α phase) takes place upon temperature variations (thermo-tropism). The transition temperature T_c depends on properties of the hydro-philic headgroup (charge, size, conformation, water binding capacity) as well as of the hydrophobic moiety (number, length, and degree of saturation of the acyl chains). A transition between the different phase states is usually accompanied by changes in the geometry of the molecules involved, i.e. of the geometrical cross-sections of the molecules and may, therefore, have an impact on the structure of the supramolecular assemblies.

We and others have found a very characteristic dependence of T_c on the chemical structure of the sugar region of LPS [20, 23, 51]. Thus, for enterobacterial strains, T_c was highest for free lipid A (around 45 °C), lowest for deep rough mutant LPS (around 30 °C), and, with increasing length of the polysaccharide portion towards completion of the O-chain (wild-type LPS), it increased again up to 37 to 40 °C. The phase transition temperatures of various nonenterobacterial LPS and lipid A were found to be significantly lower than those of *Enterobacteriaceae* [51]. This observation correlates with the fact that the former contain shorter acyl chains with a higher degree of unsaturation [8]. The addition of divalent cations or a lowering of pH caused a significant rigidification of the acyl chains of free lipid A and LPS prepara-tions and, partly, also led to an increase in T_c.

Molecular Modelling, Conformational Studies, and Crystallographic Data

In theoretical model calculations of the conformation of the heptaacyl lipid A component from *S. minnesota*, Labischinski et al. [16] found an angle of approximately 45° between the bisphosphoryl disaccharide backbone and the fatty acids chains. The fatty acids occupy positions lying almost exactly on a hexagonal lattice and the terminal methyl groups do not form a plane perpendicular to the membrane. The authors arrived at a length of 2.6 nm and a cross-section of 0.6 to 0.8 nm for the smaller and 1.2 to 1.6 nm for the longer side of the rectangular cross-section of the acyl chains. For hexaacyl Re-LPS from *E. coli*, Kastowsky et al. [52] found a number of slightly different conformations, and anisotropic lateral dimensions of 1.0–1.1 nm and 1.7–2.0 nm for the longer and shorter side, respectively. The orientation of the Kdo was found to be sensitive to the charge state of the molecule, and some free hydroxyl groups of the R-(3)-hydroxymyristoyl chains were shown to cause acyl chain packing perturbations within the hydrophobic domain, which is hexagonally densely packed. The tilt of the diglucosamine was determined to be $53 \pm 7°$ with respect to the membrane normal, with the reducing side of the diglucosamine (the 1-phosphate) emerging in the hydrophobic moiety of the neighboring molecule, and the 4'-phosphate sticking out into the aqeous phase. In contrast to these computer calculations, we showed by IR spectroscopy that the 1-phosphate is surrounded by water and the 4'-phosphate buried in the hydrophobic region close to the backbone, probably facing the 3-hydroxyl groups of the neighboring molecules [46]. The existence of this more unfavorable packing of the acyl chains was supported by the observations that several lipid A analogs and partial structures do not have a hexagonal dense packing. Furthermore, strongly reduced ΔH_c-values of lipid A and LPS as compared to saturated phospholipids with the same acyl chain length (based on the number of methylene groups) indicates a reduced packing density of the former. Modelling of the three-dimensional structure and conformational flexibility of a complete wild-type LPS [53] suggested that the Ra-LPS moiety will have an approximately cylindrical shape. Within this structure the O-specific chain was found to be the most flexible portion, preferring bent conformations and partially lying flat on top of the headgroup of neighboring molecules. The dimensions of the calculated model were in reasonable agreement with X-ray diffraction data of several dried LPS structures.

Kato et al. [54, 55] successfully crystallized synthetic lipid A and rough mutant Re- to Ra-LPS, but the crystallization of natural lipid A and wild-type LPS failed. For hexaacyl Re- and Ra-LPS, the anisotropic cross-section of the acyl chains was determined to be $1.67 \times 0.924 \text{ nm}^2$ and the length of the lipid A part to be 2.96 nm. The acyl chains of synthetic *E. coli*-type lipid A – viewed in the direction of the membrane normal – were found to form a hexagonal lattice with a lattice constant of 0.462 nm. The longitudinal axis was 4.93 nm corresponding to the bilayer thickness. The monolayer thickness derived from this value (2.47 nm) is thus slightly lower than the length of the lipid A molecule given by Labischinski et al. [16].

To overcome the restrictions of the molecular modelling studies and of the experiments with crystallized endotoxins, which result from neglecting the influence of water molecules, Obst et al. [56] performed molecular dynamics simulations of fully hydrated part structures of Re-LPS from *E. coli* in the presence and absence of Ca^{2+}. They found the cations to be located between the carboxylate and phosphate groups leading to a rigidification of the headgroup and an alteration of the conformation of the backbone, thus influencing the structure and flexibility of the hydrophobic region as well. Aggregation of these Re-LPS monomers should lead to lamellar structures. However, the authors point out that neglecting intermolecular binding between neighboring Re-LPS molecules makes a direct comparison to experimental results derived from data on aggregated lipids impossible.

A comparison of the data from various investigations on the molecular geometry of lipid A show very similar values for the bilayer thickness in the absence of water or at low water content. These values range from 4.8 to 5.2 nm and swell to a value of 5.5 nm at high water content [25]. The respective values for Re-LPS are 5.7 to 6.4 nm [21, 54], for Ra-LPS 7.8 to 8.8 nm, and the other rough mutant LPS lie in between [27]. For smooth form LPS, the bilayer lengths are reported to be 7–12 nm at low water [16] and 24–36 nm at high water content [Brandenburg and Seydel, unpubl. data] indicating a dramatic swelling of the bilayer periodicity due to the hydration of the O-chain. Thus, under these conditions the O-antigen part of the endotoxin is far from being 'heavily coiled' as found at low water content [16].

Interaction of Endotoxins with Target Cell Membranes

As stated above, there are two types of interaction of endotoxins with cell membranes of the immune system. The first is nonspecific and is driven solely by hydrophobic interactions. The second is specific and requires the aid of serum proteins like LBP. Evidence for the existence of the nonspecific interaction was found in experiments on the intercalation of different endotoxins into phospholipid liposomes or various mammalian cells (macrophages, erythrocytes and others [6, 29, 57]). However, most of these experiments were performed with the fluorescence probe diphenylhexatriene (DPH) or electron spin resonance probes. As has been shown at least for the former probe, such noncovalently-bound labels may exchange between two lipid systems, thus mimicking fusion or intercalation [58]. Fluorescence resonance energy transfer (FRET) measurements on the interaction of endotoxins with phospholipid liposomes of different composition have shown that endotoxin aggregates do not intercalate into the liposomes on a short-time scale (<0.5 h); however, after long-term incubation at 37 °C such intercalation can take place [58; and our own unpubl. results]. Experimental evidence for binding of endotoxins to monocytes/macrophages was found in the presence of serum [59]. Serum, however, contains many different proteins, a fact which hampers the study of the role of individual proteins such as LBP. From experiments with radiolabeled endotoxins, a two-step mechanism, reversible adsorption followed by irreversible membrane-binding, was proposed [60], which is in accordance with findings that specific as well as nonspecific binding of endotoxins to peritoneal macrophages is observed [59, 61]. Investigations into the LBP-mediated interaction of various endotoxins with phospholipid liposomes resembling the composition of macrophage cytoplasmic membranes demonstrated the existence of an LBP-mediated transport of endotoxin into the liposomes in a concentration-dependent manner, and the highest transport efficiency was found for the endotoxins with shortest sugar chains (lipid A, Re-LPS) [31]. Interestingly, negatively charged phospholipids could also be shown to be transported by LBP.

Relation of Physicochemical Parameters to Biological Activity

There are some striking correlations between physicochemical parameters, the phase transition temperatures of endotoxins and the molecular conformations deduced from the supramolecular aggregate structures, with responses in biological systems.

Thus, various biological effects can be correlated with the value of T_c or the state of order of the acyl chains at 37 °C. For example, the induction of monokine secretion by LPS. The strong cation-induced decrease of fluidity of LPS at 37 °C has been correlated to an enhanced monokine secretion (TNFα and IL-1β) of monocytes [62]. The increasing order of the hydrocarbon chains by Zn^{2+} could facilitate a stronger bond between LPS and LBP thus enhancing the transport of LPS to the target membrane (see also an earlier review [12]).

A most striking correlation was found between the biological activity of lipid A from different bacterial species and their preference for different conformations: Lipid A samples adopting a cylindrical conformation like those from *R. capsulatus, C. violaceum,* and *R. fulvum* were completely inactive, those assuming slightly conical conformations had intermediate activity, and those lipid A prefering conical conformations were highly active [34, 48, 50]. From these observations, the expression of a conical conformation of the lipid A moiety was deduced as prerequisite for endotoxicity ('endotoxic conformation').

In this context, the consequences of the structural characteristics of the synthetic lipid A analogs and partial structures (see above) for their bioactivity can also be understood: all samples with less than six acyl chains are almost inactive [47], which is again apparently correlated with their preference for cylindrical conformation.

As stated earlier, the lipid A phosphate groups – or possibly other negatively charged groups at the lipid A backbone – are of great importance for biological activity. We very recently showed that phosphate groups strongly influence the conformation of lipid A and with that its ability to induce cytokines in PMBC [34] (table 1). Thus, the phosphate-free synthetic lipid A (503) has a cylindrical conformation and is biologically completely inactive. The dephosphorylated Re-LPS is also cylindrical with a very slight tendency to a conical conformation, and shows IL-6 induction only at very high LPS concentrations. We have also demonstrated that a lipid A analog with the 1-phosphate substituted by a carboxymethyl group had an IL-6-inducing capacity comparable to that of the native lipid A.

Table 1. Correlation of IL-6 induction, number of negative charges, molecular conformation, and LBP-mediated transport of endotoxins into phospholipid liposomes whose composition corresponds to that of the macrophage membrane (adopted from Schromm et al. [34]).

Sample	Amount of endotoxin necessary to induce 1 ng/ml IL-6 (ng/ml)		Number of negative molecular charges	Molecular conformation	LBP-mediated transport into PL$_{M\Phi}$
	in whole blood	in PBMC			
LPS Re of *E.coli*	0.02	0.04	4		++
Dephosphorylated LPS Re of *E.coli*	12	15	2		++
Lipid A from LPS Re of *E.coli*	0.4	15	2		++
CM-506	n.d.	10	2		++
Monophosphoryl lipid A from LPS Re of *E.coli*	1.2	n.d.	1		+
Lipid A from LPS of *Rs.fulvum*	40	n.d.	1		–
Phosphate-free lipid A (synthetic compound '503')	>>1000	>>1000	0		–

Possible Activation Mechanisms or How Can the Correlation between Molecular Conformation and Biological Activity Be Understood?

Membrane-bound mCD14 is GPI-anchored to the membrane [63] and, thus, is lacking a transmembrane domain. Therefore, this protein is not capable of transmitting a signal to the cellular interior, and it has been suggested that in addition to a receptor-mediated activation the insertion of endotoxin molecules into the lipid matrix may also be a prerequisite for the activation of host cells [64]. This could be achieved via direct intercalation by hydrophobic interaction of existing endotoxin monomers (see CMC values presented above) or via the intercalation of monomers or smaller endotoxin aggregates, produced by the disaggregating properties of LBP and transported to the membrane through a complex interplay of sCD14, mCD14, and LBP [65]. As mentioned above, LBP, initially defined as lipopolysaccharide-binding protein, is not LPS-specific but rather interacts with and transports other negatively charged lipids. Thus, LBP seems to be a lipid transfer protein in a more general sense. For signaling, the mere intercalation of endotoxin molecules into the lipid matrix would not be sufficient, and two principally different fates for the molecule are conceivable. One possible mechanism is the internalization of endotoxin, which is described in several investigations [66, 67], and which could be either associated with cell activation or with clearance and detoxification. The other mechanism constitutes the direct or CD14-mediated interaction of endotoxin with other membrane-associated or membrane-spanning proteins [67], and several candidates have been proposed. One of these is an LPS-binding protein found on human monocytes and endothelial cells, which was shown to be CD55 [68]. This protein binds LPS and lipid A only in the presence of soluble CD14 and LBP. In addition, chemical cross-linking experiments revealed an association of sCD14/LPS complexes with a 216-kD membrane protein [69]. Other candidates include the purinergic receptor P2X7 [70] and ion channels [71]. Very recently, TLR2, a member of the family of Toll-like receptors, has been shown to be essential in LPS-mediated activation of cells [72, 73]. The association of LPS with TLR2 is augmented by LBP and dependent on CD14. It appears that TLR2 is responsible for transmitting an LPS-initiated signal to the cellular interior resulting in NF-κB activation and the production of endogenous mediators. However, up to now there is no experimental proof that this is the universal pathway of endotoxin-induced cytokine induction.

From the finding of a correlation between the conformation of endotoxin molecules and their bioactivity, we propose a mode of cell activation that proceeds from the existence of a transmembrane signal transducing protein, which is triggered by the binding of endotoxin molecules. The triggering signal requires a particular conformation of the lipid A component of endotoxin. Thus, only those endotoxins will be active, which possess a lipid A portion having a conical molecular conformation. We, furthermore, postulate that for binding to the signaling protein, the existence of a sufficient number of hydroxy fatty acids in lipid A is necessary to facilitate the formation of hydrogen bonds. This conformational concept would readily explain the antagonistic action of biologically inactive endotoxins. In these cases, the binding sites of the transmembrane protein would be occupied by the inactive molecules, thus inhibiting the interaction of the active structures. The fact that compound 406 (synthetic lipid A precursor Ia) and E5531 (synthetic compound related to lipid A of *R. capsulatus*) express antagonistic activity in humans, but the former not in mice and hamsters and the latter not in hamsters [10, 74], could be explained in this model by assuming that the conformation and/or the binding sites of the membrane-spanning signaling molecule or other components of the signaling cascade vary in the different systems.

In our model, the above-mentioned types of endotoxin intercalation should express different efficiencies. Thus, the direct intercalation of monomers and the LBP-mediated process will lead to a random intercalation in the lipid matrix, whereas the mCD14-mediated process will transfer endotoxin directly into the vicinity of the signaling protein assuming that mCD14 and the signal transducer protein are neighbors. This assumption is backed by the observation that mCD14-mediated activation can be blocked by anti-CD14 antibodies. At high endotoxin concentration the blockade by anti-CD14 antibodies can be overcome [75], and obviously in that case the CD14-independent activation pathway is operative.

So far, nothing is known about the processes leading to the triggering of the subsequent biochemical machinery. In the case that the transmembrane protein was an ion channel, the interaction of endotoxin with the channel protein could lead to mismatches in intracellular ion concentrations provoking the activation of the subsequent signaling cascade.

Acknowledgments

Part of the studies reported in this review were supported by the Deutsche Forschungsgemeinschaft (SFB 367, project B8; SFB 470, project B5; GRK 288, project A3) and by the Federal Minister of Education, Science, Research, and Technology (BMBF grant 01 KI 9851/A6).

References

1   Lüderitz O, Freudenberg MA, Galanos C, Lehmann V, Rietschel ET, Shaw DH: Lipopolysaccharides of gram-negative bacteria. Curr Top Membr Transport 1982;17:79–134.
2   Wilkinson SG: Bacterial lipopolysaccharides – Themes and variations. Prog Lipid Res 1997;35: 283–343.
3   Nikaido H, Vaara M: Molecular basis of bacterial outer membrane permeability. Microbial Rev 1985;49:1–32.
4   Rietschel ET, Brade H, Holst O, Brade L, Müller-Loennies S, Mamat U, Zähringer U, Beckmann F, Seydel U, Brandenburg K, Ulmer AJ, Mattern T, Heine H, Schletter J, Hauschildt S, Loppnow H, Schönbeck U, Flad H-D, Schade UF, Di Padova F, Kusumoto, S, Schumann RR: Bacterial endotoxin: Chemical constitution, biological recognition, host response, and immunological detoxification. Curr Top Microbiol Immunol 1996;216:39–81.
5   Morrison DC, Ryan JL: Bacterial endotoxins and host immune response. Adv Immunol 1979;28: 293–450.
6   Jackson SK, James PE, Rowlands CC, Mile B: Binding of endotoxin to macrophages: Interactions of spin-labelled saccharide residues. Biochim Biophys Acta 1992;1135:165–170.
7   Dijkstra J, Bron R, Wilschut J, de Haan A, Ryan JL: Activation of murine lymphocytes by lipopolysaccharide incorporated in fusogenic, reconstituted influenza virus envelopes (virosomes). J Immunol 1996;157:1028–1036.
8   Mayer H, Krauss JH, Yokota A, Weckesser J: Natural variants of lipid A; in Friedman H, Klein TW, Nakano M, Nowotny A (eds): Endotoxin. New York, Plenum Press, 1990, pp 45–70.
9   Loppnow H, Libby P, Freudenberg MA, Kraus JH, Weckesser J, Mayer H: Cytokine induction by lipopolysaccharide (LPS) corresponds to the lethal toxicity and is inhibited by nontoxic *Rhodobacter capsulatus* LPS. Infect Immun 1990;58:3743–3750.

10   Golenbock DT, Hampton RY, Qureshi N, Takayama K, Raetz CRH: Lipid A-like molecules that antagonize the effects of endotoxins on human monocytes. J Biol Chem 1991;266:19490–19498.

11   Israelachvili JN: Intermolecular and Surface Forces, ed 2. London, Academic Press, 1991, 366 pp.

12   Seydel U, Wiese A, Schromm AB, Brandenburg K: A biophysical view on function and activity of endotoxins; in Morrison D, Brade H, Opal S, Vogel S (eds): Endotoxin in Health and Disease. New York, Marcel Dekker, 1998, pp 195–220.

13   Shands JW: The physical structure of bacterial lipopolysaccharides; in Weinbaum G, Kadis S, Ajl SJ (eds): Microbial Toxins, vol IV: Bacterial Endotoxins. New York, Academic Press, 1971, pp 127–144.

14   Emmerling G, Henning U, Gulik-Krzywicki T: Order-disorder conformational transition of hydrocarbon chains in lipopolysaccharides. Eur J Biochem 1977;78:503–509.

15   Wawra H, Buschmann H, Formanek H, Formanek S: Strukturuntersuchung mit Röntgenbeugungsmethoden an Lipopolysacchariden von *Salmonella minnesota* Mutanten S SF1111 und R SF 1167. Z Naturforsch 1979;34C:171–178.

16   Labischinski H, Barnickel G, Bradaczek H, Naumann D, Rietschel ET, Giesbrecht P: High state of order of isolated bacterial lipopolysaccharide and its possible contribution to the permeation barrier property of the outer membrane. J Bacteriol 1985;162:9–20.

17   Labischinski H, Naumann D, Schultz C, Kusumoto S, Shiba T, Rietschel ET, Giesbrecht P: Comparative X-ray and Fourier-transform-infrared investigations of conformational properties of bacterial and synthetic lipid A of *Escherichia coli* and *Salmonella minnesota* as well as partial structures and analogues thereof. Eur J Biochem 1989;179:659–665.

18   Labischinski H, Vorgel E, Uebach W, May RP, Bradaczek H: Architecture of bacterial lipid A in solution. A neutron small-angle scattering study. Eur J Biochem 1990;190:359–363.

19   Hayter JB, Rivera M, McGroarty EJ: Neutron scattering analysis of bacterial lipopolysaccharide phase structure. J Biol Chem 1987;262:5100–5105.

20   Brandenburg K, Seydel U: Physical aspects of structure and function of membranes made from lipopolysaccharides and free lipid A. Biochim Biophys Acta 1984;775225–238.

21   Brandenburg K, Koch MHJ, Seydel U: Phase diagram of deep rough mutant lipopolysaccharide from *Salmonella minnesota* R595. J Struct Biol 1992;108:93–106.

22   Seydel U, Brandenburg K, Koch MHJ, Rietschel ET: Supramolecular structure of lipopolysaccharide and free lipid A under physiological conditions as determined by synchroton small-angle X-ray diffraction. Eur J Biochem 1989;186:325–332.

23   Naumann D, Schultz C, Born J, Labischinski H, Brandenburg K, von Busse G, Brade H, Seydel U: Investigations into the polymorphism of lipid A from lipopolysaccharides of *Escherichia coli* and *Salmonella minnesota* by Fourier-transform infrared spectroscopy. Eur J Biochem 1987;164: 159–169.

24   Brandenburg K, Blume A: Investigations into the thermotropic phase behaviour of natural membranes extracted from gram-negative bacteria and artificial membrane systems made from lipopolysaccharides and free lipid A. Thermochim Acta 1987;119:127–142.

25   Brandenburg K, Koch MHJ, Seydel U: Phase diagram of lipid A from *Salmonella minnesota* and *Escherichia coli* rough mutant lipopolysaccharide. J Struct Biol 1990;105:11–21.

26   Seydel U, Labischinski H, Kastowsky M, Brandenburg K: Phase behaviour, supramolecular structure, and molecular conformation of lipopolysaccharide. Immunobiol 1993;187:191–211.

27   Seydel U, Koch MHJ, Brandenburg K: Structural polymorphisms of rough mutant lipopolysaccharides Rd to Ra from *Salmonella minnesota*. J Struct Biol 1993;110:232–243.

28   Morrison, DC: Nonspecific interactions of bacterial lipopolysaccharides with membranes and membrane components; in Berry LJ (ed): Cellular Biology of Endotoxins. Amsterdam, Elsevier, 1985, pp 25–55.

29   Portoles MT, Pagani R, Diaz-Laviada I, Municio AM: Effect of *Escherichia coli* lipopolysaccharide on the microviscosity of liver plasma membranes and hepatocytes suspensions and monolayers. Cell Biochem Function 1987;5:55–61.

30   Mathison JC, Tobias PS, Wolfson E, Ulevitch RJ: Plasma lipopolysaccharide (LPS)-binding protein: A key component in macrophage recognition of gram-negative LPS. J Immunol 1992;149:200–206.

31   Schromm AB, Brandenburg K, Rietschel ET, Flad H-D, Carroll SF, Seydel U: Lipopolysaccharide binding protein (LBP) mediates CD14-independent intercalation of lipopolysaccharide into phospholipid membranes. FEBS Lett 1996;399:267–271.

32   Tobias PS, Ulevitch RJ: Lipopolysaccharide binding protein and CD14 in LPS dependent macrophage activation. Immunobiology 1993;187:227–232.

33   Kirkland TN, Finley F, Leturcq D, Moriarty A, Lee J-D, Ulevitch RJ, Tobias PS: Analysis of lipopolysaccharide binding by CD14. J Biol Chem 1993;268:24818–24823.

34   Schromm AB, Brandenburg K, Loppnow H, Zähringer U, Rietschel ET, Carroll SF, Koch MHJ, Kusumoto S, Seydel U: The charge of endotoxin molecules influences their conformation and interleukin-6 inducing capacity. J Immunol 1998;161:5464–5471.

35   Zähringer U, Lindner B, Rietschel ET: Molecular structure of lipid A, the endotoxic center of bacterial lipopolysaccharides. Adv Carbohydr Chem Biochem 1994;50:211–276.

36   Rau H, Seydel U, Freudenberg M, Weckesser J, Mayer H: Lipopolysaccharide of the phototrophic bacterium *Rhodospirillum fulvum*. System Appl Microbiol 1995;18:154–163.

37   Rietschel ET, Kirikae T, Loppnow H, Zabel P, Ulmer J, Brade H, Seydel U, Zähringer U, Schlaak M, Flad H-D, Schade U: Molecular aspects of the chemistry and biology of endotoxin; in Sies H, Flohé L, Zimmer G (eds): Molecular Aspects of Inflammation. Berlin, Springer, 1991, pp 207–231.

38   TakadaH, Kotani S: Structural requirements of lipid A for endotoxicity and other biological activities. CRC Crit Rev Microbiol 1989;16:477–523.

39   Mattern T, Thanhäuser A, Reiling N, Toellner K-M, Duchrow M, Kusumoto S, Rietschel ET, Ernst M, Brade H, Flad H-D, Ulmer AJ: Endotoxin and lipid A stimulate proliferation of human T cells in the presence of autologous monocytes. J Immunol 1994;153:2996–3004.

40   Wang M-H, Feist W, Herzbeck H, Brade H, Kusumoto S, Rietschel ET, Flad H-D, Ulmer AJ: Suppressive effect of lipid A partial structures on lipopolysaccharide or lipid A-induced release of interleucin 1 by human monocytes. FEMS Microbiol Immunol 1990;64:179–186.

41   Kirikae T, Schade FU, Kirikae F, Qureshi N, Takayama K, Rietschel ET: Diphosphoryl lipid A derived from the lipopolysaccharide (LPS) of *Rhodobacter sphaeroides* ATCC 17023 is a potent competitive LPS inhibitor in murine macrophage-like J774.1 cells. FEMS Immunol Med Microbiol 1994;9:237–244.

42   Hofer M, Hampton RY, Raetz CRH, Yu H: Aggregation behavior of lipid IVa in aqueous solutions at physiological pH. 1. Simple buffer solutions. Chem Phys Lipids 1991;59:167–181

43   Takayama K, Din ZZ, Mukerjee P, Cooke PH, Kirkland TN: Physicochemical properties of the lipopolysaccharide unit that activates B lymphocytes. J Biol Chem 1990;265:14023–14029.

44   Din ZZ, Mukerjee P, Kastowsky M, Takayama K: Effect of pH on solubility and ionic state of lipopolysaccharide obtained from the deep rough mutant of *Escherichia coli*. Biochemistry 1993; 32:4579–4586.

45   Naumann D, Schultz C, Sabisch A, Kastowsky M, Labischinski H: New insights into the phase behaviour of a complex anionic amphiphile: Architecture and dynamics of bacterial deep rough mutant lipopolysaccharide membranes as seen by FTIR, X-ray, and molecular modelling techniques. J Molec Struct 1989;214:213–246.

46   Brandenburg K, Kusumoto S, Seydel U: Conformational studies of synthetic lipid A analogues and partial structures by infrared spectroscopy. Biochim Biophys Acta 1997;1329:193–201.

47   Brandenburg K, Richter W, Koch MHJ, Meyer HW, Seydel U: Characterization of the nonlamellar cubic and H$_{II}$ structures of lipid A from *Salmonella enterica* serovar Minnesota by X-ray diffraction and freeze-fracture electron microscopy. Chem Phys Lipids 1998;91:53–69.

48   Brandenburg K, Mayer H, Koch MHJ, Weckesser J, Rietschel ET, Seydel U: Influence of the supramolecular structure of free lipid A on its biological activity. Eur J Biochem 1993;218:555–563.

49   Seydel U, Brandenburg K, Rietschel ET: A case for an endotoxic conformation. Prog Clin Biol Res 1994;388:17–30.

50   Brandenburg K, Seydel U, Schromm AB, Loppnow H, Koch MHJ, Rietschel ET: Conformation of lipid A, the endotoxic center of bacterial lipopolysaccharide. J Endotoxin Res 1996;3:173–178.

51   Brandenburg K, Seydel U: Investigation into the fluidity of lipopolysaccharide and free lipid A membrane systems by Fourier-transform infrared spectroscopy and differential scanning calorimetry. Eur J Biochem 1990;191:229–236.

52 Kastowsky M, Sabisch A, Gutberlet T, Bradaczek H: Molecular modelling of bacterial deep rough mutant lipopolysaccharide of *Escherichia coli*. Eur J Biochem 1991;197:707–716.

53 Kastowsky M, Gutberlet T, Bradaczek H: Molecular modelling of the three-dimensional structure and conformational flexibility of bacterial lipopolysaccharide. J Bacteriol 1992;174:4798–4806.

54 Kato N, Ohta M, Kido N, Ito H, Naito S, Hasegawa T, Watabe T, Sasaki K: Crystallization of R-form lipopolysaccharides from *Salmonella minnesota* and *Escherichia coli*. J Bacteriol 1990;172: 1516-1528.

55 Kato N, Ohta M, Kido N, Arakawa Y, Sugiyama T, Naito S, Ito H: Polymorphism of crystals of *Salmonella minnesota* Re and Ra lipopolysaccharides. Microbiol Immunol 1993;37:549–555.

56 Obst S, Kastowsky M, Bradaczek H: Molecular dynamics simulations of six different fully hydrated monomeric conformers of *Escherichia coli* Re-lipopolysaccharide in the presence and absence of Ca^{2+}. Biophys J 1997;72:1031–1046.

57 Larsen NE, Enelow RI, Simons ER, Sullivan R: Effect of bacterial endotoxin on the transmembrane electrical potential and plasma membranes of human monocytes. Biochim Biophys Acta 1985;815:1–8.

58 Schromm AB, Brandenburg K, Rietschel ET, Seydel U: Do endotoxin aggregates intercalate into phospholipid membranes in a nonspecific, hydrophobic manner? J Endotoxin Res 1995;2:313–323.

59 Tahri-Jouti MA, Chaby R: Binding of endotoxin to macrophages: Distinct effects of serum constituents. Immunol Invest 1991;20:377–386.

60 Carr C, Morrison DC: A two-step mechanism for the interaction of Re lipopolysaccharide with erythrocyte membranes. Rev Infect Dis 1984;6:497–500.

61 Tahri-Jouti M-A, Chaby R: Specific binding of lipopolysaccharide to mouse macrophages. I. Characteristics of the interaction and inefficiency of the polysaccharide region. Mol Immunol 1990;27: 751–761.

62 Wellinghausen N, Schromm AB, Seydel U, Brandenburg K, Luhm J, Kirchner H, Rink L: Zinc enhances lipopolysaccharide-induced monokine secretion by alteration of fluidity state of lipopolysaccharide. J Immunol 1996;157:3139–3145.

63 Haziot A, Chen S, Ferrero E, Low MG, Silber R, Goyert SM: The monocyte differentiation antigen, CD14, is anchored to the cell membrane by a phosphatidylinositol linkage. J Immunol 1988;141: 547–552.

64 Morrison DC, Rudbach JA: Endotoxin-cell-membrane interactions leading to transmembrane signaling. Contemp Top Mol Immunol 1981;8:187–218.

65 Wurfel MM, Wright SD: Lipopolysaccharide-binding protein and soluble CD14 transfer lipopolysaccharide to phospholipid bilayers: Preferential interaction with particular classes of lipid. J Immunol 1997;158:3925–3934.

66 Kriegsmann J, Gay S, Bräuer R: Endocytosis of lipopolysaccharide in mouse macrophages. Cell Mol Biol 1993;39:791–800.

67 Thiéblemont N, Thieringer R, Wright S: Innate immune recognition of bacterial lipopolysaccharide: Dependence on interactions with membrane lipids and endocytic movement. Immunity 1998;8: 771–777.

68 Schletter J, Brade H, Brade L, Krüger C, Loppnow H, Kusumoto S, Rietschel ET, Flad H-D, Ulmer AJ: Binding of lipopolysaccharide (LPS) to an 80-kilodalton membrane protein of human cells is mediated by soluble CD14 and LPS-binding protein. Infect Immun 1995;63:2576–2580.

69 Vita N, Lefort S, Sozzani P, Reeb R, Richards R, Borysiewicz LK, Ferrara P, Labéta MO: Detection and biochemical characteristics of the receptor for complexes of soluble CD14 and bacterial lipopolysaccharide. J Immunol 1997;158:3457–3462.

70 Hu Y, Fisette PL, Denlinger LC, Guadarrama AG, Sommer JA, Proctor RA, Bertics PJ: Purinergic receptor modulation of lipopolysaccharide signaling and inducible nitric-oxide synthase expression in RAW 264.7 macrophages. J Biol Chem 1998;273:27170–27175.

71 Maruyama N, Yasunori K, Yamauchi K, Aizawa T, Ohrui T, Nara M, Oshiro T, Ohno L, Tanura G, Shimura S, Saschi H, Tahishima T, Shirato K: Quinine inhibits production of tumor necrosis factor-α from human alveolar macrophages. Am J Respir Cell Mol Biol 1994;10:514–520.

72 Yang RB, Mark MR, Gray A, Huang A, Xie MH, Zhang M, Goddard A, Wood WI, Gurney AL, Godowski PJ: Toll-like receptor-2 mediates lipopolysaccharide-induced cellular signaling. Nature 1998;395:284–288.

73 Kirschning CJ, Wesche H, Ayers M, Rothe M: Human Toll-like receptor 2 confers responsiveness to bacterial LPS. J Exp Med 1999;188:2091–2097.
74 Delude RL, Savedra R Jr, Zhao H, Thieringer R, Yamamoto S, Fenton MJ, Golenbock DT: CD14 enhances cellular responses to endotoxin without imparting ligand-specific recognition. Proc Natl Acad Sci USA 1995;92:9288–9292.
75 Lynn WA, Liu Y, Golenbock DT: Neither CD14 nor serum is absolutely necessary for activation of mononuclear phagocytes by bacterial lipopolysaccharide. Infect Immun 1993;61:4452–4461.

Prof. Dr. U. Seydel, Research Center Borstel, Center for Medicine and Biosciences,
Division of Biophysics, Parkallee 10, D–23845 Borstel (Germany)
Tel. +49 4537 188 232, Fax +49 4537 188 632, E-Mail useydel@fz-borstel.de

Jack RS (ed): CD14 in the Inflammatory Response.
Chem Immunol. Basel, Karger, 2000, vol 74, pp 25–41

....

Structure/Function Relationships of CD14

Felix Stelter

Institut für Immunologie und Transfusionsmedizin, Ernst-Moritz-Arndt-Universität
Greifswald, Deutschland

The mammalian immune system has developed mechanisms to generate an almost infinite number of receptor molecules (T-cell receptors and antibodies) which provide the organism the potential to specifically recognize almost any antigenic determinant on any pathogenic microorganism. This diversity comes at a price: killing of pathogens by antibodies or effector T cells involves proliferation of antigen-specific T- and B-cell clones, a process which requires several days or even weeks to complete. Therefore, the immediate and often life-saving immune response against an invading pathogen relies on a more or less nonspecific recognition of microbial constituents by a limited array of pattern recognition receptors. These receptors target structures which share a certain degree of similarity in different strains of microorganisms, i.e. chemical patterns commonly found on pathogens (e.g. lipopolysaccharide (LPS), lipoteichoic acid, muramyldipeptide) but not present on host tissues. A number of pattern recognition receptors are currently known including the scavenger receptors, the family of Toll-like receptors and the protein which is the focus of the present review: CD14 [1].

Molecular Characteristics of CD14

The CD14 molecule is preferentially expressed on monocytes and some tissue macrophages and to a much lesser extent on polymorphonuclear granulocytes [2, 3]. In mice, Fearns et al. [4] found extramyeloid expression of CD14 mRNA in many tissues with the highest levels in uterus, adipose tissue, and lung, which could be markedly increased by exposure to LPS. In addition, Diamond et al. [5] described CD14 expression in tracheal epithelial cells and Liu et al. [6] detected CD14 protein in hepatocytes from LPS-treated mice.

The human CD14-cDNA was cloned in 1988 by Ferrero and Goyert [7, 8] and the gene has been mapped to chromosome 5 into a region (5q 23-31) which codes for growth factors and receptors. The gene is organized in a simple fashion: it consists of only two exons which are separated by an 88-bp intron. The first exon only codes for the first amino acid methionine. The genomic structure does not provide any evidence of exon coded domains or alternative splicing mechanisms.

The human protein consists of 375 amino acid (aa) residues from which a 19 aa signal peptide is cleaved in the endoplasmatic reticulum [9]. The sequence ends with a 21 aa hydrophobic domain and lacks the characteristic basic residues of a stop transfer domain. Haziot et al. [9] and Simmons et al. [10] demonstrated that the protein is attached to the cell surface by a glycosyl-phosphatidyl-inositol (GPI) anchor and can be removed by phophatidylinosi-tol-specific phospholipase C treatment. CD14 has four potential N-linked glycosylation sites (Asn-X-Thr/Ser, X not Pro) and, in addition, bears O-linked carbohydrates [11]. The carbohydrate content accounts for about 20% of the total molecular mass of the mature 53-kD glycoprotein [12]. Although total lack of N-linked carbohydrates is associated with diminished CD14 surface expression on transfected CHO cells [11], glycosylation has not yet been linked to a functional property of the molecule [11, 13].

The primary sequence of human CD14 is highly homologous (approximately 65%) to its mouse [14], rat [15], rabbit [16] and bovine [17] counterparts. Besides this CD14 shows significant homology only to an inducible gram-negative bacteria-binding protein from the silkworm [18]. CD14 contains 10 leucine-rich repeats located between residues 86 and 329 (human) [19]. Leucine-rich repeats are found in a number of phylogenetically distant proteins and are thought to mediate protein-protein interactions [20]. However, no function has yet been ascribed to them in CD14.

Functional Properties of CD14

Wright et al. [21] first discovered that CD14 is a key receptor on the surface of monocytes required for the induction of an inflammatory response triggered by pico- and nanomolar amounts of endotoxin [22–24]. However, recognition of LPS by CD14 is not straightforward: it needs to be catalyzed by lipopolysaccharide-binding protein (LBP), an acute-phase serum reactant [21, 25–27]. The binding stoichiometry LPS:CD14 has been reported to be

1:1 [28] and Kirkland et al. [29] determined a K_d value of 27 nM. Because LPS tends to aggregate in solution, stoichiometric data are conflicting: under different experimental conditions Gegner et al. [30] observed the formation of large ternary complexes consisting of LPS, LBP and CD14.

A substantial fraction of the organisms CD14 is present as a soluble form (sCD14) in serum and body fluids. As a soluble LPS receptor sCD14 may inhibit LPS-induced activation of monocytes by competition with the membrane-bound molecule: Schütt and coworkers found that excess of sCD14 blocks LPS binding to monocytes [31] and the oxidative burst response of mononuclear cells [32]. In line with these results Haziot et al. [33] showed an inhibition of the LPS-stimulated TNFα production in human whole blood cultures. However, sCD14 may not only act as a neutralizing agent but also as a stimulatory soluble receptor which facilitates the LPS-induced activation of cells lacking membrane-bound CD14 (mCD14). A variety of reports have demonstrated that lack of mCD14 on monocytes derived from patients suffering from paroxysmal nocturnal hemoglobinuria can be compensated for by sCD14 [34–36] and, similarly, that activation of CD14-negative endothelial, epithelial and smooth muscle cells by LPS is mediated through an sCD14-dependent pathway [37–43].

In 1994, Pugin et al. [44] brought up the idea of CD14 as a pattern recognition element by demonstrating binding of CD14 to cell wall preparations of gram-positive bacteria and to lipoarabinomannan from *Mycobacterium tuberculosis*. Till now, several studies have confirmed interactions of CD14 with a variety of ligands derived from different microbial sources and even with synthetic molecules. These ligands include cell wall constituents from *Staphylococcus aureus* [45, 46], soluble peptidoglycan [47–49], a streptococcal neutral polysaccharide [50], lipoproteins and lipopeptides from *Treponema pallidum* and *Borrelia burgdorferi* [51], a peptide derived from the WI-1 protein of *Blastomyces dermatidis* [52] and polymannuronic acid [53]. In addition to this we could show that CD14 does not only recognize soluble molecules but also whole gram-negative *E. coli* bacteria [54, 55] and even the gram-positive species *Bacillus subtilis* [56]. Apart from CD14 as a pattern recognition receptor, Yu et al. [57] proposed that the protein may play a broader physiological role in lipid exchange mechanisms: the authors demonstrated that sCD14 catalyzes the exchange of different phospholipids including phosphatidylinositol, phosphatidylcholine, and phosphatidylethanolamine. In line with this Devitt et al. [58] found that CD14 on the surface of human macrophages is important for the recognition and clearance of apoptotic cells.

Generation of Soluble CD14 and Soluble CD14 Isoforms

In healthy subjects sCD14 is present at a concentration of 1–6 µg/ml [59]. Depending on the method employed different isoform patterns of sCD14 can be detected. Initially, two forms, a high-molecular-mass form of 53–56 kD and a low-molecular-mass form of 48–50 kD which are separated on SDS-PAGE have been described in human serum [12, 60]. In addition, by isoelectric focusing six forms with isoelectric points in the range of pH 4.5–5.2 are distinguished [12]. In supernatants of MonoMac-6 cells Labeta et al. [61] found at least 7 isoforms by 2D electrophoresis.

Increased levels of soluble CD14 have been reported in a number of clinical situations. These situations include chronic inflammatory diseases such as psoriasis [62], sarcoidosis [63] and systemic lupus erythematosus [64] and chronic infections such as HIV infection [65, 66]. The molecule has also been in focus as a putative surrogate marker for the monitoring of intensive care patients: Krüger et al. [67] detected elevated sCD14-serum levels in patients suffering from polytrauma and burns and Endo et al. [68] found higher sCD14 levels in septic patients with multi-organ failure. In line with this Landmann et al. [69] correlated sCD14 values with increased mortality in gram-negative septic shock and demonstrated the predominance of a high-molecular-mass (55 kD) isoform in patients with high levels of sCD14. Furthermore, Martin et al. [70] observed an increase of sCD14 in sera and bronchoalveolar lavage fluids of patients with acute respiratory distress syndrome. Although the role of sCD14 in these situations is not yet understood it appears that sCD14 is a marker of chronic and acute inflammation and may be viewed as an acute phase protein. Therefore tremendous effort has been made to study mechanisms involved in the generation of soluble CD14 and soluble CD14 isoforms.

Bazil and Strominger [71] first reported a release of CD14 from the surface of monocytes after stimulation with various activating agents including phorbol 12-myristate 13-acetate, interferon-γ, LPS, a calcium ionophore and anti-CD14 antibodies. The authors further demonstrated that the size of sCD14 shed from the monocyte surface is smaller than that of either the membrane-bound form or a soluble CD14 cleaved by treatment with phosphatidylinositol-specific phospholipase C. Downmodulation of membrane CD14 was totally blocked at 4 °C or at pH 4.5 and markedly inhibited by protease inhibitors. Thus, it appears that generation of sCD14 by activated monocytes is facilitated by an inducible protease. Later, Bufler et al. [72] provided evidence that sCD14 may also be produced by a protease-independent mechanism: structural analysis of sCD14 released by transfected cell lines confirmed the presence of the C-terminus predicted from the primary cDNA sequence. Since GPI anchor attachment is associated with the removal of a hydrophobic C-terminal signal peptide, this indicates that sCD14 may escape posttranslational modification and can be directly secreted. This pathway has in vivo relevance because identical results were obtained with sCD14 derived from peritoneal dialysis fluid of a patient with kidney dysfunction. Durieux et al. [73] showed that the high-molecular-mass form of sCD14 is released faster than the low-molecular-mass form. They demonstrated that a fraction of the low-molecular-mass form is converted from mCD14 by a pathway involving endocytosis followed by exocytosis. Absence of the eight C-terminal amino acids prevented mCD14 expression but not the secretion of a high-molecular-mass form of sCD14.

In our study [11], we extended these findings by comparing the isoform patterns on SDS gels of sCD14 purified from human serum or from culture media of CHO cells expressing either wild-type CD14 or a truncated mutant, $CD14_{(1-335)}$, which lacks the GPI attachment signal. Two major (53 and 50 kD) and two minor (46 and 43 kD) isoforms were detected in serum and in CHO-supernatants independently of whether the cells were transfected with the truncated or the full-length cDNA. This indicates that different forms of sCD14 may be released directly without first passing through an mCD14 stage. However, after enzymatic removal of carbohydrates $sCD14_{(1-335)}$ is distinct from the wild-type protein: its isoform pattern seems to be completely due to differences in N-linked glycosylation. In contrast, after deglycosylation of proteins isolated from serum or from CHO cells transfected with a wild-type construct two forms of apparent molecular masses of 44 and 41 kD remain detectable. It appears that the 41 kD-form moves faster than the deglycosylation product seen with $sCD14_{(1-335)}$. Because $sCD14_{(1-335)}$ cannot be expressed on the cell surface this suggests that conversion of mCD14 into sCD14 is associated with proteolytic cleavage resulting in a protein smaller than $sCD14_{(1-335)}$. These results show that the isoform pattern of sCD14 detectable by SDS-PAGE is due both to different extents of N-linked glycosylation and to a second mechanism which appears to involve proteolytic cleavage.

The possible mechanisms of sCD14 secretion are summarized in figure 1. It is not yet clear to what extent these pathways contribute to the serum sCD14 pool and how certain mechanisms are regulated during inflammatory conditions.

Identification of the LPS-Binding Site of Human CD14

Several groups have focused on identifying the LPS-binding domain of human CD14. Viriyakosol and Kirkland [74] constructed a series of single and combined mutants by deleting the small hydrophilic regions at positions

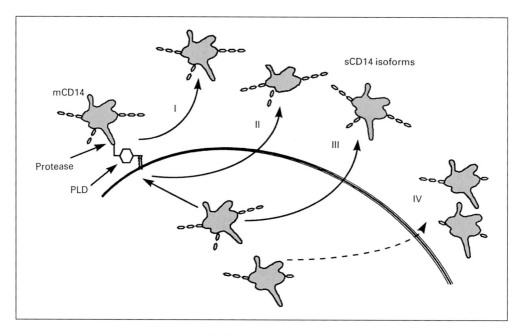

Fig. 1. Pathways of sCD14 release. I = sCD14 may be cleaved from the cell surface by a protease [71] or alternatively by a cell-associated phospholipase D (PLD) as demonstrated for other GPI-linked molecules [106]. II = Generation of a low-molecular-mass form of sCD14 by a process involving endocytosis followed by exocytosis [73]. III = The synthesis of a GPI anchor may be leaky and therefore a fraction of the sCD14 is directly secreted. IV = Minor forms of a lower molecular weight are not completely glycosylated [11].

9–12, 15–18, 35–39 and 59–63. The authors could not detect direct binding of LPS to any of their mutants expressed on CHO cells. In a functional assay cells transfected with mutant proteins responded less efficiently to LPS by translocation of NFκB but only cells expressing a deletion mutant spanning residues 35–39 were entirely unresponsive. This study indicated that hydrophilic regions within the 65 N-terminal amino acids of CD14 are critical for serum-dependent binding of LPS to CD14.

Juan et al. [75] sequentially deleted the leucine-rich repeats of human CD14 and tested the functional properties of soluble CD14 fragments derived from supernatants of transfected cells. The shortest fragment which could be secreted by COS-7 cells contains three leucine-rich repeats and consists of 152 aa. This fragment, $sCD14_{(1-152)}$, binds LPS and enables cellular responses to LPS. The authors concluded that the N-terminal 152 residues of human CD14 confer essentially wild-type bioactivity in vitro. A similar result was obtained by Viriyakosol and Kirkland [76] who generated a chimeric protein consisting of $CD14_{(1-151)}$ fused to the C-terminal region of decay-accelerating factor (DAF). The membrane-bound chimera expressed on the surface of CHO- and 70Z/3 cells was found to be a fully functional LPS receptor in both cell lines.

In a further study McGinley et al. [13] discovered that the region 57 to 64 is protected by LPS from the action of Asp-N protease and chymotrypsin. In an accompanying report this result was corroborated by Juan et al. [77] who demonstrated that a deletion of this particular region results in a nonfunctional mutant. A soluble mutant, $sCD14_{Δ57-64}$, neither binds LPS directly in a native PAGE assay nor does it trigger cellular responses to LPS. Taken together both results suggested that region 57 to 64 is the LPS-binding site of human CD14.

However, this conclusion was not confirmed by our study [78] in which a panel of alanine substitution mutants comprising the entire N-terminal 152 aa region of human CD14 was analyzed. In each mutant a block of five amino acids, with the exception of cysteines and prolines, was substituted by alanine. Also N-linked glycosylation sites were spared because it has been shown that glycosylation has no role in LPS-binding [11, 13]. Altogether 23 mutant cDNAs were stably transfected into CHO cells. Four of 21 expressed mutants, $CD14_{(9-13)A}$, $CD14_{(39-41, 43-44)A}$, $CD14_{(51-55)A}$ and $CD14_{(57, 59,61-63)A}$, are not recognized by certain 'inhibitory' anti-CD14 monoclonal antibodies [79] which interfere with the binding of LPS to human monocytes thus indicating a spatial relation of epitopes 9–13, 39–44 and 51–63 to the LPS-binding site. However, whereas mutants $CD14_{(9-13)A}$, $CD14_{(51-55)A}$ and $CD14_{(57, 59, 61-63)A}$ bind LPS with almost identical K_d values as compared to the wild-type protein (26.1, 28.8, 26.4 vs. 31.9, respectively) only $CD14_{(39-41, 43-44)A}$ is unable to react with FITC-labelled and tritiated LPS or to mediate LPS-induced translocation of NFκB in transfected CHO cells. This demonstrates that the integrity of region 39–44 is crucial for serum-dependent binding of LPS to human CD14.

Furthermore, our study shows that alanine substitutions in region 57–64 do not alter the functional properties of CD14, suggesting that residues located within this region do not contribute to the chemical interaction between CD14 and LPS. The latter result is a discrepancy to the study of Juan et al. [77] and is likely to be due to a different mutagenesis strategy. By deleting 8 aa Juan et al. [77] may have removed a 'space holder' in the protein and may have caused a more severe alteration of the tertiary structure.

Nevertheless, the data of McGinley et al. [13] strongly point to region 57–64 as part of a conformational epitope required to interact with LPS: a bound LPS molecule is probably located in close proximity to this site because otherwise it would not protect from the action of proteases. In addition, LPS binding can be blocked by the monoclonal antibody MEM 18 which recognizes an epitope in region 51–64 [77, 78]. Besides this we found evidence for a third region comprising the acidic residues (Asp-Asp-Glu-Asp-Phe) at position 9–13 which is structurally associated with the LPS-binding site. First, the 'inhibitory' antibodies biG 10–13 which do not recognize $CD14_{(39-41, 43-44)A}$ also do not react with $CD14_{(9-13)A}$. Second, serum-dependent binding of FITC-labelled LPS to a combined mutant, $CD14_{(9-13/57, 59, 61-63)A}$ is hardly detectable [80]. Determination of the Kd value reveals that the combined mutant has a fourfold higher Kd (124 nM) than measured for $CD14_{(9-13)A}$, and $CD14_{(57, 59, 61-63)A}$. Together, both regions 9–13 and 57–64 might contribute to a three-dimensional structure which forms the LPS-binding groove of human CD14.

A common feature of other LPS-binding molecules such as LBP, bactericidal permeability-increasing protein, Limulus anti-LPS factor and polymyxin B is the presence of positively charged amino acids which are thought to interact with the negatively charged phosphate groups of the lipid A moiety of LPS [81–85]. Currently available data suggest that the LPS-binding domain of CD14 does not share this trait. In this regard it is important to note that Porro et al. [86] have used features of LPS-binding peptide derivatives from polymyxin B to select for candidate lipid A binding sequences in the primary structure of LPS-binding proteins from phylogenetically distinct species including human and mouse CD14. They found that a nonapeptide spanning aa 67–75 of human CD14 (Val-Lys-Ala-Leu-Arg-Val-Arg-Arg-Leu) was able to inhibit the clotting reaction in the LAL assay induced by LPS derived from six different gram-negative bacterial species. Furthermore, the peptide was active in vivo in inhibition of a local hemorrhagic necrosis reaction after intradermal injection of LPS into the skin of New Zealand white rabbits. The prominent feature of this peptide is the presence of four basic, positively charged amino acids. This raises the question of whether region 67–75 of human CD14 does also participate in LPS binding. Two mutants from our panel, $CD14_{(65-68, 70)A}$ and $CD14_{(71-75)A}$, are located within this region [78]. Whereas $CD14_{(65-68, 70)A}$ is perfectly able to bind LPS, we could not detect expression of $CD14_{(71-75)A}$, either as a membrane-bound or as a soluble protein, in two independently transfected cell lines. We have therefore generated two separate mutants, $CD14_{(71, 72)A}$ and $CD14_{(73-75)A}$. Both mutants are expressed on the cell surface of CHO cells and bind LPS with similar affinities as compared to wild-type CD14 (Kd 34.1 and 35.7 nM, respectively) [Menzel and Stelter, unpubl. observ.]. Thus, although a peptide derived from this region is able to interact with LPS, it appears that the basic region between aa 67–75 is not essential for recognition of LPS by human CD14.

As outlined above, CD14 does not only bind soluble LPS but also whole bacteria and a variety of ligands derived from different biological sources. This raises the question of whether these molecules interact with the same binding site on CD14. We have shown that the integrity of region 39–44 is important for recognition of both gram-negative *E. coli* and gram-positive *B. subtilis* bacteria [56, 78]. In addition, CD14$_{(39-41, 43-44)A}$ does not form stable complexes with lipoteichoic acid derived from *B. subtilis* [56]. Kusunoki et al. [45] found that deletion of residues 57–64, which eliminates LPS binding also abrogates CD14-dependent responses to *S. aureus* molecules. Finally, by using monoclonal antibodies Dziarski et al. [87] demonstrated that conformational binding sites on CD14 for soluble peptidoglycan and LPS are partially identical and partially different. Together, these data suggest that the domain by which different molecules are recognized by CD14 is at least similar if not identical. However, the common pattern shared by all these ligands and how it fits into the binding pocket of CD14 has not yet been defined. Further studies using X-ray crystallography or NMR techniques are required to finally solve this problem.

CD14-Induced Cellular Signalling

Because CD14 lacks a transmembrane domain it has long been a question of dispute how after binding of LPS a signal is delivered into the cell. Lee et al. [88] transfected different CD14-fusion proteins into 70Z/3 cells. These constructs either contained a GPI anchor or transmembrane domains derived from tissue factor or from a MHC class I molecule. It was shown that GPI-anchored or integral membrane forms of CD14 mediate identical cellular responses to LPS including NFκB activation and p38 tyrosine phosphorylation. Furthermore, the anti-CD14 monoclonal antibody 18E12 did not block binding of LPS to CD14 but did inhibit LPS-induced cellular activation. These results indicate that binding of endotoxin to membrane-bound CD14 is followed by subsequent interactions with additional membrane component(s) which enable transmembrane signalling. It is conceivable that a similar mechanism might be responsible for the activation of CD14-negative cells in the presence of LPS complexed with soluble CD14. Thus, recent research has focused on three principal topics: characterizing domains of CD14 required for interaction with the signal transducer, identifying putative coreceptor molecules, and determining whether sCD14- and mCD14-dependent pathways of cellular activation are different.

As outlined above Juan et al. [75] have shown that a CD14 fragment consisting of the N-terminal 152 aa is a fully functional LPS receptor, i.e. the fragment is able to mediate LPS-induced cellular signalling. This indicates that the domain of human CD14 required for association with a putative coreceptor must be located within this region. In a further study the same group [89] generated an sCD14 mutant containing alanine substitutions in the region between aa 7-10. sCD14$_{(7-10)A}$ has an impaired ability to mediate

LPS-dependent IL-6 production and NFκB activation in the astrocytoma cell line U 373. Furthermore, integrin activation in neutrophils by LPS is decreased in the presence of sCD14$_{(7-10)A}$ as compared to wild-type sCD14. Thus, region 7–10 of human CD14 is necessary for LPS-induced cellular activation but not for LPS binding. These results were extended by our study in which we have tested the soluble proteins derived from the supernatants of our CHO cells transfected with a panel of alanine substitution mutants (see above). Eighteen of the transfected cell lines release soluble CD14 into the culture supernatant. We analyzed the capacity of serum-free supernatants containing mutant proteins to facilitate LPS-induced IL-6 production in U 373 astrocytoma cells [90]. Three mutants, sCD14$_{(9-13)A}$, sCD14$_{(91-94, 96)A}$, and sCD14$_{(97-101)A}$, failed to activate U 373 cells in this assay. The substitutions in CD14$_{(9-13)A}$ and in the mutant constructed by Juan et al. [89] fall into the same N-terminal region. Besides this it appears that a second region between amino acids 91–101 has also a role in the signalling of LPS. To determine the contribution of these two regions to LPS-induced cellular activation it would be of great interest to examine in detail how CD14 (and CD14 mutants) interact with a coreceptor molecule on the surface of cells.

Till now, a number of molecules have been brought into the discussion as putative signal tranducers: Schletter et al. [91] isolated an 80-kD protein from membrane preparations of the human monocytic cell line Mono-Mac-6. In the presence of serum this protein was able to interact with smooth- and rough-type LPS, and also with lipid A. The serum factors required to mediate binding of lipid A to the 80-kD protein were identified as sCD14 and LBP. Recently, the protein has been shown to be identical to DAF/CD55 [92]. Because DAF is a GPI anchored protein present on many cell types including erythrocytes the functional significance of these observations remains to be established. Zarewych et al. [93] provided evidence for a role of complement receptor 3 (CD11b/CD18; CR3) in CD14-mediated signalling. Using resonance energy transfer microscopy, it was found that in the presence of serum or LBP LPS triggers the association of CD14 with CR3 on the surface of neutrophils. CR3 might indeed be another intermediate in the signalling chain: Ingalls et al. [94] showed that CR3 expressed on CHO cells facilitates LPS induced NFκB translocation presumably by presenting LPS to a downstream signal transducer. Vita et al. [95] detected specific and saturable binding of iodinated sCD14 to a 216-kD binding structure on several cell types, including monocytes, endothelial and epithelial cells. Specific binding to all cell types occurred only when both sCD14 and LPS were present. Furthermore, the authors provided evidence that the sCD14-LPS receptor may also interact with mCD14. However, the primary structure of the 216-kD protein has not yet been elucidated. Very recently, Yang et al. [96] demonstrated that Toll-like receptor 2 (TLR2) is a signalling receptor that is activated by LPS in a response that depends on LBP and is enhanced by CD14. A similar but not identical result was obtained by Kirschning et al. [97], who found that coexpression of CD14 stimulated LPS signal transmission through TLR2. In addition, purified soluble CD14 could substitute for serum to support LPS-induced TLR2 activation. In the mouse system another Toll-like receptor, TLR4, has been found to be involved in LPS signalling: In two independent reports Poltorak et al. [98] and Qureshi [99] discovered that mutations of TLR4 selectively impede LPS signal transduction in endotoxin-resistant C3H/HeJ and C57BL/10ScCr mice. These results have been corroborated by Hoshino et al. [100] who demonstrated that macrophages and B cells from TLR4-deficient mice do not respond to LPS and Chow et al. [101] who found that LPS-sCD14 complexes stimulate NFκB-mediated gene expression in HEK 293 cells transfected with the TLR4 gene.

In contrast to these reports which employ a coreceptor for LPS-signalling Joseph et al. [102] proposed an alternative pathway: after integration of LPS into the membrane it mimics the second messenger function of ceramide and stimulates cells by interaction with a ceramide-activated kinase. In line with this result Wurfel and Wright [103] showed that sCD14 transfers LPS into phospholipid bilayers.

Altogether, although a considerable amount of data has been accumulated, it is not yet clear to what extend all these proposed signal transducing pathways contribute to LPS-induced, CD14-mediated cellular activation in different cells. Furthermore, the signalling cascade initiated by LPS binding to mCD14 may be distinct from that switched on by LPS-sCD14 complexes in CD14-negative cells. In their initial study, Lee et al. [88] showed that sCD14 cannot substitute for mCD14 in the activation of 70Z/3 cells. From our studies [78, 80, 90] it appears that membrane-bound $CD14_{(9-13)A}$ is able to transmit a signal into the cell, because CHO cells expressing this mutant respond to LPS by translocation of NFκB whereas the soluble mutant, $sCD14_{(9-13)A}$, is defective in mediating the activation of U 373 cells. Haziot et al. [104] described two anti-CD14 antibodies which block the LPS-sCD14-induced activation of endothelial cells but do not affect the response of monocytes to LPS. Taken together these data suggest that the putative signal transducer which mediates the activation of mCD14-bearing cells is indeed distinct from that required for LPS-sCD14-induced stimulation of cells.

Concluding Remarks

CD14 is thought to play a major role in recognition of microbial pathogens and has been shown to be a key protein in septic shock induced by LPS and

E. coli [105]. Therefore, tremendous effort has been invested in collecting data on CD14 structure and function. Still, the number of different pieces do not yield a complete puzzle. None of the questions touched in this review are finally solved. Several pathways have been implicated in the generation of soluble CD14 but how do all these pathways relate to the in vivo situation? Data derived from mutational analyses have proven the importance of certain regions of the molecule for recognition of LPS and for CD14-mediated signalling but the three-dimensional structure has not yet been determined. Also, what is the common pattern shared by the different ligands which have been shown to interact with CD14? Different molecules have been discussed as putative link in the LPS-CD14 signalling chain but which molecule is the real signal transducer in a given cell?

References

1   Medzhitov R, Janeway CA Jr: Innate immunity: The virtues of a nonclonal system of recognition. Cell 1997;91:295–298.
2   Goyert SM, Ferrero E: Biochemical analysis of myeloid antigens and cDNA expression of gp 55 (CD14); in McMichael A (ed): Leukocyte Typing III. Oxford, Oxford University Press, 1987, pp 613–619.
3   Matsuura K, Ishida T, Setoguchi M, Higuchi Y, Akizuki S, Yamamoto S: Upregulation of mouse CD14 expression in Kupffer cells by lipopolysaccharide. J Exp Med 1994;179:1671–1676.
4   Fearns C, Kravchenko VV, Ulevitch RJ, Loskutoff DJ: Murine CD14 gene expression in vivo: Extramyeloid synthesis and regulation by lipopolysaccharide. J Exp Med 1995;181:857–866.
5   Diamond G, Russell JP, Bevins CL: Inducible expression of an antibiotic peptide gene in lipopoly-saccharide-challenged tracheal epithelial cells. Proc Natl Acad Sci USA 1996;93:5156–5160.
6   Liu S, Khemlani LS, Shapiro RA, Johnson ML, Liu K, Geller DA, Watkins SC, Goyert SM, Billiar TR: Expression of CD14 by hepatocytes: Upregulation by cytokines during endotoxemia. Infect Immun 1998;66:5089–5098.
7   Ferrero E, Goyert SM: Nucleotide sequence of the gene encoding the monocyte differentiation antigen, CD14. Nucl Acids Res 1988;16:4173.
8   Goyert SM, Ferrero E, Rettig WJ, Yenamandra AK, Obata F, Le Beau MM: The CD14 monocyte differentiation antigen maps to a region encoding growth factors and receptors. Science 1988;239: 497–500.
9   Haziot A, Chen S, Ferrero E, Low MG, Silber R, Goyert SM: The monocyte differentiation antigen, CD14, is anchored to the cell membrane by a phosphatidylinositol linkage. J Immunol 1988;141: 547–552.
10   Simmons DL, Tan S, Tenen DG, Nicholson-Weller A, Seed B: Monocyte antigen CD14 is a phospholipid anchored membrane protein. Blood 1989;73:284–289.
11   Stelter F, Pfister M, Bernheiden M, Jack RS, Bufler P, Engelmann H, Schütt C: CD14 is N- and O-glycosylated: Contribution of N-linked glycosylation to different sCD14-isoforms. Eur J Biochem 1996;236:457–464.
12   Bazil V, Horejsi V, Baudys M, Kristofova H, Strominger JL, Kostka W, Hilgert I: Biochemical characterisation of a soluble form of the 53-kDa monocyte surface antigen. Eur J Immunol 1986; 16:1583–1589.
13   McGinley MD, Narhi LO, Kelley MJ, Davy E, Robinson J, Rohde MF, Wright SD, Lichenstein HS: CD14: Physical properties and identification of an exposed site that is protected by lipopoly-saccharide. J Biol Chem 1995;270:5213–5218.

14   Setoguchi M, Nasu N, Yoshida S, Higuchi Y, Akizuki S, Yamamoto S: Mouse and human CD14 (myeloid cell-specific leucine-rich glycoprotein) primary structure deduced from cDNA clones. Biochim Biophys Acta 1989;1008:213–222.

15   Takai N, Kataoka M, Higuchi Y, Matsuura K, Yamamoto S: Primary structure of rat CD14 and characteristics of rat CD14, cytokine, and NO synthase mRNA expression in mononuclear phagocyte system cells in response to LPS. J Leukoc Biol 1997;61:736–744.

16   Tobias PS, Mathison J, Mintz D, Lee JD, Kravchenko V, Kato K, Pugin J, Ulevitch RJ: Participation of lipopolysaccharide-binding protein in lipopolysaccharide-dependent macrophage activation. Am J Respir Cell Mol Biol 1992;7:239–245.

17   Ikeda A, Takata M, Taniguchi T, Sekikawa K: Molecular cloning of bovine CD14 gene. J Vet Med Sci 1997;59:715–719.

18   Lee WJ, Lee JD, Kravchenko VV, Ulevitch RJ, Brey PT: Purification and molecular cloning of an inducible gram-negative bacteria-binding protein from the silkworm, *Bombyx mori*. Proc Natl Acad Sci USA 1996,93:7888–7893.

19   Ferrero E, Hsieh CL, Francke U, Goyert SM: CD14 is a member of the family of leucine-rich proteins and is encoded by a gene syntenic with multiple receptor genes. J Immunol 1990:145: 331–336.

20   Kobe B, Deisenhofer J: The leucine-rich repeat: A versatile binding motif. Trends Biochem Sci 1994; 19:415–421.

21   Wright SD, Ramos RA, Tobias PS, Ulevitch RJ, Mathison JC: CD14, a receptor for complexes of lipopolysaccharide (LPS) and LPS binding protein. Science 1990;249:1431–1433.

22   Couturier C, Haeffner-Cavaillon N, Caroff M, Kazatchkine MD: Binding sites for endotoxins (lipopolysaccharides) on human monocytes. J Immunol 1991;147:1899–1904.

23   Lee JD, Kato K, Tobias PS, Kirkland TN, Ulevitch RJ: Transfection of CD14 into 70Z/3 cells dramatically enhances the sensitivity to complexes of lipopolysaccharide (LPS) and LPS binding protein. J Exp Med 1992;175:1697–1705.

24   Dentener MA, Bazil V, von Asmuth EJ, Ceska M, Buurman WA: Involvement of CD14 in lipopoly-saccharide-induced tumor necrosis factor-alpha, IL-6 and IL-8 release by human monocytes and alveolar macrophages. J Immunol 1993;150:2885–2891.

25   Mathison JC, Tobias PS, Wolfson E, Ulevitch RJ: Plasma lipopolysaccharide (LPS)-binding protein. A key component in macrophage recognition of gram-negative LPS. J Immunol 1992;149:200–206.

26   Martin TR, Mathison JC, Tobias PS, Leturcq DJ, Moriarty AM, Maunder RJ, Ulevitch RJ: Lipopolysaccharide binding protein enhances the responsiveness of alveolar macrophages to bacterial lipopolysaccharide: Implications for cytokine production in normal and injured lungs. J Clin Invest 1992;90:2209–2219.

27   Hailman, E, Lichenstein HS, Wurfel MM, Miller DS, Johnson DA, Kelley M, Busse LA, Zukowski MM, Wright SD: Lipopolysaccharide (LPS)-binding protein accelerates the binding of LPS to CD14. J Exp Med 1994;179:269–277.

28   Kitchens RL, Munford RS: Enzymatically deacylated lipopolysaccharide (LPS) can antagonise LPS at multiple sites in the LPS recognition pathway. J Biol Chem 1995;270:9904–9910.

29   Kirkland TN, Finley F, Leturcq D, Moriarty A, Lee JD, Ulevitch RJ, Tobias PS: Analysis of lipopolysaccharide binding by CD14. J Biol Chem 1993;268:24818–24823.

30   Gegner JA, Ulevitch RJ, Mathison JC: Lipopolysaccharide (LPS) signal transduction and clearance. Dual roles for LPS binding protein and membrane CD14. J Biol Chem 1995;270:5320–5325.

31   Grunwald U, Krüger C, Schütt C: Endotoxin-neutralising capacity of soluble CD14 is a highly conserved specific function. Circ Shock 1993;39:220–225.

32   Schütt C, Schilling T, Grunwald U, Schönfeld W, Krüger C: Endotoxin neutralising capacity of soluble CD14. Res Immunol 1992;143:71–78.

33   Haziot A, Rong GW, Bazil V, Silver J, Goyert SM: Recombinant soluble CD14 inhibits LPS-induced tumor necrosis factor-a production by cells in whole blood. J Immunol 1994;152:5868–5876.

34   Duchow J, Marchant A, Crusiaux A, Husson C, Alonso-Vega C, De-Groote D, Neve P, Goldman M: Impaired phagocyte responses to lipopolysaccharide in paroxysmal nocturnal hemoglobinuria. Infect Immun 1993;61:4280–4285.

35 Sundan A, Ryan L, Brinch L, Espevik T, Waage A: The involvement of CD14 in stimulation of TNF production from peripheral mononuclear cells isolated from PNH patients. Scand J Immunol 1995;41:603–608.

36 Schütt C, Schilling T, Grunwald U, Stelter F, Witt S, Krüger C, Jack RS: Human monocytes lacking the membrane-bound form of the bacterial lipopolysaccharide (LPS) receptor CD14 can mount an LPS-induced oxidative burst response mediated by a soluble form of CD14. Res Immunol 1995; 146:339–350.

37 Frey EA, Miller DS, Gullstein Jahr T, Sundan A, Bazil V, Espevik T, Finlay BB, Wright SD: Soluble CD14 participates in the response of cells to lipopolysaccharide. J Exp Med 1992;176:1665–1671.

38 Pugin J, Schürer-Maly CC, Leturcq D, Moriarty A, Ulevitch RJ, Tobias PS: Lipopolysaccharide activation of human endothelial and epithelial cells is mediated by lipopolysaccharide-binding protein and soluble CD14. Proc Natl Acad Sci USA 1993;90:2744–2748.

39 Read MA, Cordle SR, Veach RA, Carlisle CD, Hawiger J: Cell-free pool of CD14 mediates activation of transcription factor NF-kappa B by lipopolysaccharide in human endothelial cells. Proc Natl Acad Sci USA 1993;90: 9887–9891.

40 Haziot A, Rong GW, Silver J, Goyert SM: Recombinant soluble CD14 mediates the activation of endothelial cells by lipopolysaccharide. J Immunol 1993;151:1500–1507.

41 Arditi M, Zhou J, Dorio R, Rong GW, Goyert SM, Kim KS: Endotoxin-mediated endothelial cell injury and activation: Role of soluble CD14. Infect Immun 1993;61:3149–3156.

42 Goldblum SE, Brann TW, Ding X, Pugin J, Tobias PS: Lipopolysaccharide (LPS)-binding protein and soluble CD14 function as accessory molecules for LPS-induced changes in endothelial barrier function, in vitro. J Clin Invest 1994;93:692–702.

43 Loppnow H, Stelter F, Schönbeck U, Schlüter C, Ernst M, Schütt C, Flad HD: Endotoxin activates human smooth muscle cells despite lack of expression of CD14 mRNA or endogenous membrane CD14. Infect Immun 1995;63:1020–1026.

44 Pugin J, Heumann D, Tomasz A, Kravchenko VV, Akamatsu Y, Nishijima M, Glauser MP, Tobias PS, Ulevitch RJ: CD14 is a pattern recognition receptor. Immunity 1994;1:509–516.

45 Kusunoki T, Hailman E, Juan TS, Lichenstein HS, Wright SD: Molecules from *Staphylococcus aureus* that bind CD14 and stimulate innate immune responses. J Exp Med 1995;182:1673–1682.

46 Kusunoki T, Wright SD: Chemical characteristics of *Staphylococcus aureus* molecules that have CD14-dependent cell-stimulating activity. J Immunol 1996;157:5112–5117.

47 Weidemann B, Brade H, Rietschel ET, Dziarski R, Bazil V, Kusumoto S, Flad HD, Ulmer AJ: Soluble peptidoglycan-induced monokine production can be blocked by anti-CD14 monoclonal antibodies and by lipid A partial structures. Infect Immun 1994;62:4709–4715.

48 Gupta D, Kirkland TN, Viriyakosol S, Dziarski R: CD14 is a cell-activating receptor for bacterial peptidoglycan. J Biol Chem 1996;271:23310–23316.

49 Weidemann B, Schletter J, Dziarski R, Kusumoto S, Stelter F, Rietschel ET, Flad HD, Ulmer AJ: Specific binding of soluble peptidoglycan and muramyldipeptide to CD14 on human monocytes. Infect Immun 1997;65:858–864.

50 Soell M, Lett E, Holveck F, Schöller M, Wachsmann D, Klein JP: Activation of human monocytes by streptococcal rhamnose glucose polymers is mediated by CD14 antigen, and mannan binding protein inhibits TNF-α release. J Immunol 1995;154:851–860.

51 Sellati TJ, Bouis DA, Kitchens RL, Darveau RP, Pugin J, Ulevitch RJ, Gangloff SC, Goyert SM, Norgard MV, Radolf JD: *Treponema pallidum* and *Borrelia burgdorferi* lipoproteins and synthetic lipopeptides activate monocytic cells via a CD14-dependent pathway distinct from that used by lipopolysaccharide. J Immunol 1998;160:5455–5464.

52 Newman SL, Chaturvedi S, Klein BS: The WI-1 antigen of *Blastomyces dermatidis* yeasts mediates binding to human macrophage CD11b/CD18 (CR3) and CD14. J Immunol 1995;154:753–761.

53 Espevik T, Otterlei M, Skjåk-Bræk G, Ryan L, Wright SD, Sundan A: The involvement of CD14 in stimulation of cytokine production by uronic acid polymers. Eur J Immunol 1993;23: 255–261.

54 Jack RS, Grunwald U, Stelter F, Workalemahu G, Schütt C: Both membrane-bound and soluble forms of CD14 bind to gram-negative bacteria. Eur J Immunol 1995;25: 1436–1441.

55 Grunwald U, Fan X, Jack RS, Workalemahu G, Kallies A, Stelter F, Schütt C: Monocytes can phagocytose gram-negative bacteria by a mechanism dependent on CD14. J Immunol 1996;157: 4119–4125.

56 Fan X, Stelter F, Menzel R, Jack RS, Spreitzer I, Hartung T, Schütt C: Structures in *Bacillus subtilis* are recognised by CD14 in an LBP-dependent reaction. Infect Immun 1999;67:2964–2968.

57 Yu B, Hailman E, Wright SD: Lipopolysaccharide binding protein and soluble CD14 catalyse exchange of phospholipids. J Clin Invest 1997;99:315–324.

58 Devitt A, Moffatt OD, Raykundalia C, Capra JD, Simmons DL, Gregory CD: Human CD14 mediates recognition and phagocytosis of apoptotic cells. Nature 1998;392:505–509.

59 Grunwald U, Krüger C, Westermann J, Lukowsky A, Ehlers M, Schütt C: An enzyme-linked immunosorbent assay for the quantification of solubilised CD14 in biological fluids. J Immunol Meth 1992;155:225–232.

60 Bazil V, Baudys M, Hilgert I, Stefanova I, Low MG, Zbrozek J, Horejsi V: Structural relationship between the soluble and membrane bound forms of human monocyte surface glycoprotein CD14. Mol Immunol 1989;26:657–662.

61 Labeta MO, Durieux JJ, Fernandez N, Herrmann R, Ferrara P: Release from a human monocyte-like cell line of two different soluble forms of the lipopolysaccharide receptor, CD14. Eur J Immunol 1993;23:2144–2155.

62 Schöpf RE, Dobmeyer J, Dobmeyer T, Morsches B: Soluble CD14 monocyte antigen in suction blister fluid and serum of patients with psoriasis. Dermatology 1993;186:45–49.

63 Pforte A, Schiessler A, Gais P, Beer B, Ehlers M, Schütt C, Ziegler-Heitbrock HWL: Expression of CD14 correlates with lung function impairment in pulmonary sarcoidosis. Chest 1994;105: 349–354.

64 Nockher WA, Wigand R, Schoeppe W, Scherberich JE: Elevated levels of soluble CD14 in serum of patients with systemic lupus erythematosus. Clin Exp Immunol 1994;96:15–19.

65 Nockher WA, Bergmann L, Scherberich JE: Increased soluble CD14 serum levels and altered CD14 expression of peripheral blood monocytes in HIV-infected patients. Clin Exp Immunol 1994;98:369–374.

66 Lien E, Aukrust P, Sundan A, Müller F, Froland SS, Espevik T: Elevated levels of serum-soluble CD14 in human immunodeficiency virus type 1 (HIV-1) infection: Correlation to disease progression and clinical events. Blood 1998;92:2084–2092.

67 Krüger C, Schütt C, Obertacke U, Joka T, Müller FE, Knoller J, Koller M, König W, Schönfeld W: Serum CD14 levels in polytraumatised and severely burned patients. Clin Exp Immunol 1991; 85:297–301.

68 Endo S, Inada K, Kasai T, Takakuwa T, Nakae H, Kikuchi M, Yamashita H, Yoshida M: Soluble CD14 (sCD14) levels in patients with multiple organ failure (MOF). Res Commun Chem Pathol Pharmacol 1994;84:17–25.

69 Landmann R, Zimmerli W Sansano S, Link S, Hahn A, Glauser MP, Calandra T: Increased circulating soluble CD14 is associated with high mortality in gram-negative septic shock. J Infect Dis 1995;171:639–644.

70 Martin TR, Rubenfeld GD, Ruzinski JT, Goodman RB, Steinberg KP, Leturcq DJ, Moriarty AM, Raghu G, Baughman RP, Hudson LD: Relationship between soluble CD14, lipopolysaccharide binding protein, and the alveolar inflammatory response in patients with acute respiratory distress syndrome. Am J Respir Crit Care Med 1997; 155:937–944.

71 Bazil V, Strominger JL: Shedding as a mechanism of down-modulation of CD14 on stimulated human monocytes. J Immunol 1991;147:1567–1574.

72 Bufler P, Stiegler G, Schuchmann M, Hess S, Krüger C, Stelter F, Schütt C, Eckershorn C, Engelmann H: Soluble LPS receptor (CD14) is released via two different mechanisms from human monocytes and CD14 transfectants. Eur J Immunol 1995;25:604–610.

73 Durieux JJ, Vita N, Popescu O, Guette F, Calzada-Wack J, Munker R, Schmidt RE, Lupker J, Ferrara P, Ziegler-Heitbrock HWL, Labeta MO: The two soluble forms of the lipopolysaccharide receptor, CD14: Characterisation and release by normal human monocytes. Eur J Immunol 1994; 9:2006–2012.

74 Viriyakosol S, Kirkland TN: A region of human CD14 required for lipopolysaccharide binding. J Biol Chem 1995;270:361–368.

75   Juan TSC, Kelley MJ, Johnson DA, Busse LA, Hailman E, Wright SD, Lichenstein HS: Soluble CD14 truncated at amino acid 152 binds lipopolysaccharide (LPS) and enables cellular response to LPS. J Biol Chem 1995;270:1382–1387.

76   Viriyakosol S, Kirkland TN: The N-terminal half of membrane CD14 is a functional cellular lipopolysaccharide receptor. Infect Immun 1996;64:653–656.

77   Juan TSC, Hailman E, Kelley MJ, Busse LA, Davy E, Empig CJ, Narhi LO, Wright SD, Lichenstein HS: Identification of a lipopolysaccharide binding domain in CD14 between amino acids 57 and 64. J Biol Chem 1995;270:5219–5224.

78   Stelter F, Bernheiden M, Menzel R, Jack RS, Witt S, Fan X, Pfister M, Schütt C: Mutation of amino acids 39-44 of human CD14 abrogates binding of lipopolysaccharide and Escherichia coli. Eur J Biochem 1997;243:100–109.

79   Schütt C, Witt S, Grunwald U, Stelter F, Schilling T, Fan X, Marquart BP, Bassarab S, Krüger C: Epitope mapping of CD14 glycoprotein; in Schlossmann SF, Boumsell L, Gilks W, Harlan JM, Kishimoto T, Morimoto C, Ritz J, Shaw S, Silverstein R, Springer T, Tedder TF, Todd RF (eds): Leukocyte Typing V. Oxford, Oxford University Press, 1995, pp 785–788.

80   Stelter F, Bernheiden M, Menzel R, Jack RS, Witt S, Fan X, Schütt C: Structure/function relationships of human CD14: Analysis of LPS-binding to membrane bound CD14 and LPS-induced signalling mediated by soluble CD14; in Faist E (ed): The Immune Consequences of Trauma, Shock and Sepsis – Mechanisms and Therapeutic Approaches. Bologna, Monduzzi Editore, 1997, pp 21–25.

81   Beamer LJ, Carroll SF, Eisenberg D: The BPI/LBP family of proteins: A structural analysis of conserved regions. Protein Sci 1998;7:906–914.

82   Hoess A, Watson S, Siber GR, Liddington R: Crystal structure of an endotoxin-neutralising protein from the horseshoe crab, Limulus anti-LPS factor, at 1.5 A resolution. EMBO J 1993;12:3351–3356.

83   Lamping N, Hoess A, Yu B, Park TC, Kirschning CJ, Pfeil D, Reuter D, Wright SD, Herrmann F, Schumann RR: Effects of site-directed mutagenesis of basic residues (Arg 94, Lys 95, Lys 99) of lipopolysaccharide (LPS)-binding protein on binding and transfer of LPS and subsequent immune cell activation. J Immunol 1996;157:4648–4656.

84   Schumann RR, Lamping N, Hoess A: Interchangeable endotoxin-binding domains in proteins with opposite lipopolysaccharide-dependent activities. J Immunol 1997;159:5599–5605.

85   Rustici A, Velucchi M, Faggioni R, Sironi M, Ghezzi P, Quataert S, Green B, Porro M: Molecular mapping and detoxification of the lipid A binding site by synthetic peptides. Science 1993;259: 361–365.

86   Velucchi M, Rustici A, Porro M: Molecular requirements of peptide structures binding to the lipid A region of bacterial endotoxins; in Chanock RM, Brown F, Ginsberg HS, Norrby E (eds): Vaccines 94. Cold Spring Harbor Laboratory Press, 1994, pp 141–146.

87   Dziarski R, Tapping RI, Tobias PS: Binding of bacterial peptidoglycan to CD14. J Biol Chem 1998;273:8680–8690.

88   Lee JD, Kravchenko V, Kirkland TN, Han J, Mackman N, Moriarty A, Leturcq D, Tobias PS, Ulevitch RJ: Glycosyl-phosphatidylinositol-anchored or integral membrane forms of CD14 mediate identical cellular responses to endotoxin. Proc Natl Acad Sci USA 1993;90:9930–9934.

89   Juan TSC, Hailman E, Kelley MJ, Wright SD, Lichenstein HS: Identification of a domain in soluble CD14 essential for lipopolysaccharide (LPS) signalling but not LPS binding. J Biol Chem 1995; 270:17237–17242.

90   Stelter F, Bernheiden M, Menzel R, Witt S, Jack RS, Grunwald U, Fan X, Schütt C: The molecular basis of therapeutic concepts utilising CD14. Prog Clin Biol Res 1998;397:301–313.

91   Schletter J, Brade H, Brade L, Krüger C, Loppnow H, Kusumoto S, Rietschel ET, Flad HD, Ulmer AJ: Binding of lipopolysaccharide (LPS) to an 80-kilodalton membrane protein of human cells is mediated by soluble CD14 and LPS-binding protein. Infect Immun 1995;63:2576–2580.

92   El-Samalouti VT, Schletter J, Chyla I, Lentschat A, Mamat U, Brade L, Flad HD, Ulmer AJ, Hamann L: Identification of the 80 kDa LPS-binding protein (LMP80) as decay accelerating factor (DAF, CD55). FEMS Immunol Med Microbiol 1999;23:259–264.

93   Zarewych DM, Kindzelskii AL, Todd RF 3rd, Petty HR: LPS induces CD14 association with complement receptor type 3, which is reversed by neutrophil adhesion. J Immunol 1996;156:430–433.

94 Ingalls RR, Arnaout MA, Golenbock DT: Outside-in signalling by lipopolysaccharide through a tailless integrin. J Immunol 1997;159:433–438.

95 Vita N, Lefort S, Sozzani P, Reeb R, Richards S, Borysiewicz LK, Ferrara P, Labeta MO: Detection and biochemical characteristics of the receptor for complexes of soluble CD14 and bacterial lipopolysaccharide. J Immunol 1997;158:3457–3462.

96 Yang RB, Mark MR, Gray A, Huang A, Xie MH, Zhang M, Goddard A, Wood WI, Gurney AL, Godowski PJ: Toll-like receptor-2 mediates lipopolysaccharide-induced cellular signalling. Nature 1998;395:284–288.

97 Kirschning CJ, Wesche H, Merrill Ayres T, Rothe M: Human toll-like receptor 2 confers responsiveness to bacterial lipopolysaccharide. J Exp Med 1998;188:2091–2097.

98 Poltorak A, He X, Smirnova I, Liu MY, Huffel CV, Du X, Birdwell D, Alejos E, Silva M, Galanos C, Freudenberg M, Ricciardi-Castagnoli P, Layton B, Beutler B: Defective LPS signalling in C3H/HeJ and C57BL/10ScCr mice: Mutations in Tlr4 gene. Science 1998;282:2085–2088.

99 Qureshi ST, Lariviere L, Leveque G, Clermont S, Moore KJ, Gros P, Malo D: Endotoxin-tolerant mice have mutations in Toll-like receptor 4 (Tlr4). J Exp Med 1999;189:615–625.

100 Hoshino K, Takeuchi O, Kawai T, Sanjo H, Ogawa T, Takeda Y, Takeda K, Akira S: Toll-like receptor 4 (TLR4)-deficient mice are hyporesponsive to lipopolysaccharide: Evidence for TLR4 as the Lps gene product. J Immunol 1999;162:3749–3752.

101 Chow JC, Young DW, Golenbock DT, Christ WJ, Gusovsky F: Toll-like receptor-4 mediates lipopolysaccharide-induced signal transduction. J Biol Chem 1999;274:10689–10692.

102 Joseph CK, Wright SD, Bornmann WG, Randolph JT, Kumar ER, Bittman R, Liu J, Kolesnick RN: Bacterial lipopolysaccharide has structural similarity to ceramide and stimulates ceramide-activated protein kinase in myeloid cells. J Biol Chem 1994;269:17606–17610.

103 Wurfel MM, Wright SD: Lipopolysaccharide-binding protein and soluble CD14 transfer lipopolysaccharide to phospholipid bilayers: Preferential interaction with particular classes of lipid. J Immunol 1997;158:3925–3934.

104 Haziot A, Katz I, Rong GW, Lin XY, Silver J, Goyert SM: Evidence that the receptor for soluble CD14:LPS complexes may not be the putative signal-transducing molecule associated with membrane-bound CD14. Scand J Immunol 1997;46:242–245.

105 Haziot A, Ferrero E, Kontgen F, Hijiya N, Yamamoto S, Silver J, Stewart CL, Goyert SM: Resistance to endotoxin shock and reduced dissemination of gram-negative bacteria in CD14-deficient mice. Immunity 1996;4:407–414.

106 Metz CM, Brunner G, Choi-Muira NH, Nguyen H, Gabrilove J, Caras IW, Altszuler N, Rifkin DB, Wilson EL, Davitz MA: Release of GPI-anchored membrane proteins by a cell-associated GPI-specific phospholipase D. EMBO J 1994;13:1741–1751.

Felix Stelter, Institut für Immunologie und Transfusions-medizin,
Ernst-Moritz-Arndt-Universität Greifswald,
Klinikum Sauerbruchstrasse, D–17487 Greifswald (Germany)
Tel. +49 3834 86 54 55, Fax +49 3834 86 54 90, E-Mail Stelter@uni-greifswald.de

Jack RS (ed): CD14 in the Inflammatory Response.
Chem Immunol. Basel, Karger, 2000, vol 74, pp 42–60

....

Lipopolysaccharide-Binding Protein

Ralf R. Schumann[a], *Eicke Latz*[b]

[a] Institut für Mikrobiologie und Hygiene (Head: *U.B. Göbel*, MD, PhD), and
[b] Klinik für Chirurgie und Chirurgische Onkologie, Robert-Rössle-Klinik (Head: *P.M. Schlag*, MD, PhD), Universitätsklinikum Charité, Medizinische Fakultät der Humboldt-Universität zu Berlin, Germany

Lipopolysaccharide-Binding Protein and CD14:
Their Potential Function within the Innate Immune System

The rapid and vigorous inflammatory response to toxic bacterial cell wall products, such as lipopolysaccharide (LPS) is part of the host innate immune response which provides the first line of defence against invading pathogens [1–3]. Only when this – largely phagocyte-based – defence system fails does the highly effective specific immune system come into play to protect the host against repeated invasion by pathogenic microorganisms [4, 5]. Innate immunity has attracted much attention lately and some of its molecular mechanisms have recently been elucidated [6, 7]. This cellular response of the host to bacterial components consists of a variety of direct and indirect antimicrobial mechanisms including phagocytosis, complement activation, and the release of mediators, such as cytokines, growth factors and chemokines [8–11]. Furthermore, a dysbalanced cytokine release as a result of bacterial stimulation is viewed as one major cause of sepsis and septic shock, life-threatening complications of infections [12–15]. Host molecules like CD14 and LPS-binding protein (LBP) which regulate the activation of immune cells by bacterial products such as LPS, may thus play an important role in both the initiation of innate immunity and the pathophysiology of sepsis.

Besides soluble CD14, which is thoroughly discussed in other chapters of this volume, another protein with the ability to recognize and bind LPS is found in high concentrations in serum of humans, and of every higher species investigated. This is the LBP [16, 17]. In initial experiments complexes of LBP and LPS were shown to be bound by cellular CD14 and this led to an enhanced cellular stimulation compared to that resulting from LPS alone. The function of CD14 as part of an LPS receptor was demonstrated in 1990 when purified LBP became available, and it was shown that CD14 binds complexes of LBP and LPS [18]. Although LBP and CD14 are apparently the two major endotoxin binding proteins of the host, they do not share any sequence homology, and most likely utilize different mechanisms for LPS binding. Furthermore, the mechanisms of release of the two proteins into the bloodstream are completely different. While synthesis of LBP as an acute-phase protein is regulated transcriptionally, CD14 release is mainly caused by shedding of the cell-bound receptor molecule [19, 20]. However, there are a number of similarities between LBP and CD14: both are present in similar quantities in serum; both increase in concentration during sepsis [21–23] and both have approximately the same molecular weight. For the regulation of certain LPS-dependent effects, LBP clearly synergize with CD14 – predominantly with the membrane-bound form (mCD14).

Recently, we have shown that different LBP concentrations may have distinct modulating activities on LPS dependent effects. Thus while low concentrations of LBP enhance LPS activity, the acute-phase rise in LBP concentrations inhibits LPS-induced cellular stimulation [24]. This dual activity profile is reminiscent of the situation with soluble CD14 (sCD14), which has been reported to enhance LPS effects towards CD14-negative cells, while high concentrations of this protein administered in vivo inhibit LPS activity [25–30]. For CD14, the ability to recognize cell wall compounds of other pathogenic microorganisms, such as peptidoglycan (PG) or lipoarabinomannan (LAM) has been reported earlier, leading to the suggestion that CD14 is a pattern recognition receptor [31–34]. While in earlier experiments similar results could not be obtained for LBP, current, mainly unpublished results, point in a similar direction: LBP was also found to be able to bind and mediate effects of cell wall products of gram-positive bacteria, such as lipoteichoic acids (LTA) [own unpubl. results].

Till now neither the 3-dimensional structure of CD14 nor that of LBP has been solved, nor has the mechanism of LPS binding and transfer of LPS been clearly elucidated. However, a computer-generated model for the LBP structure has been proposed recently. It is based on the crystal structure of bactericidal/permeability increasing protein (BPI), another LPS-binding protein which shares a high degree of sequence homology with LBP [35–37]. For both sCD14 and LBP, structure function analyses have been performed which identified the presumptive LPS-binding sites [38–44]. The two proteins most likely use quite different mechanisms of LPS binding, with LBP-LPS interactions relying on charge and the ability to bind to LPS multimers, while CD14 most likely predominantly binds to LPS monomers. Finally, both the genes for LBP and CD14 have been knocked-out in mice lately, resulting in animals much less responsive to LPS challenge [45–47].

LBP Is a Member of a Family of Structurally and Functionally Related Proteins

LBP is a glycosylated 58-kD protein synthesized in hepatocytes of the liver, and is released into the bloodstream upon acute-phase stimulation [16, 17, 20, 48]. Its normal concentration in human serum is 5–15 µg/l and it increases during the acute-phase response by up to 30-fold. By analysis of their genes and genomic organization it was found that LBP belongs to a family of lipid-binding proteins which includes bactericidal/permeability increasing protein (BPI), phospholipid ester transfer protein (PLTP), and cholesterol ester transfer protein (CETP) [49–55]. LBP binds to the lipid A part of bacterial lipopolysaccharide and in low concentrations catalyzes its transfer to cellular or to soluble CD14 [56–58]. Recently, LBP has been found to bind also to certain phospholipids as do other lipid binding proteins of the family, such as PLTP [59, 60]. Besides binding to LPS, LBP was also found to bind to the surface of whole bacteria and to LPS-coated erythrocytes resulting in enhanced attachment of these particles to macrophages [61]. Evidence suggests that LBP connects LPS-coated particles with macrophages by first binding LPS, then binding the particle to macrophages. This suggests that LBP has an opsonin function in vivo promoting the interaction of LBP-coated bacteria with phagocytes. With the recent discovery of members of the Toll protein family being involved in LPS signaling, it was found that LBP also enhances binding of LPS to the Toll-like receptor 2 [62–67].

LBP binds to the lipid A portion of LPS and the binding site has recently been delineated by mutagenesis experiments [40, 68]. This binding site apparently consists of a particular pattern of charged amino acid residues, which is also found in BPI and in an endotoxin-binding protein of the horseshoe crab, named ENP or LALF [43, 69]. The highest sequence homology within the LBP family is found between LBP and BPI, another LPS binding protein present in neutrophilic granulocytes [49, 70–72]. BPI also binds to LPS, however, it is unable to monomerize LPS, and clearly blocks LPS activity towards cells. The 3-dimensional structure of BPI has been solved recently, and BPI was shown to exhibit a boomerang-like shape, containing two phospholipid-binding-pockets [35]. Based on these results, a computer model for LBP has been proposed and it is very likely that the overall shape of LBP is very similar to that of BPI [36]. We have used the coordinates provided by Dr. Beamer to construct a 3-dimensional LBP model, and marked the LPS-binding region of LBP defined by mutagenesis experiments [40] (fig. 1). This LPS-binding region is located within the N-terminal half of the molecule and is exposed with several positive charged amino acids pointing away from the molecule. These amino acids display a high degree of similarity within LBP and BPI, whereas charges on the other end of the molecule differ.

Numerous in vitro experiments have shown that LBP in low concentrations can enhance LPS effects. This is most likely achieved by monomerization of LPS complexes and presentation of an LPS monomer to the CD14 receptor as outlined in more detail below [17, 73]. Constitutive LBP levels enhance LPS effects and initiate cellular responses at subthreshhold LPS levels potentially enabling the host to detect invading gram-negative bacteria and LPS early. Thus, LPS-induced TNF production and TNF-mRNA expression in rabbit peritoneal macrophages are enhanced when LPS is complexed to low-dose LBP. Similarly, rabbit peritoneal macrophages rendered unresponsive to LPS stimulation by a process called adaptation, can be restored in their ability to produce TNF by the addition of LBP [74]. Furthermore, macrophages detect and bind LPS faster when it is complexed with LBP, so that LBP acts as an opsonin for gram-negative bacteria [61, 75]. Finally, the LPS-induced responses in neutrophils can be enhanced by addition of low-dose LBP to the experimental system [76].

LBP Is Upregulated during the Acute Phase

LBP serum levels rise dramatically in the course of trauma, the 'sytemic inflammatory response syndrome' (SIRS), and sepsis [own unpubl. results]. This rise in LBP levels is caused by transcriptional activation of the LBP gene mediated by IL-1 and IL-6 [20, 48, 77, 78]. LBP is a class 1 acute-phase protein induced by IL-1 alone or synergistically by IL-1 and IL-6. Transcriptional regulation of acute-phase proteins (APRs) in the liver is based on the proximity of macrophages releasing proinflammatory cytokines upon stimulation, and hepatocytes synthesizing APRs [79]. Increase in APR biosynthesis is usually caused by increased gene transcription, mediated through cis-acting promoter elements that are binding sites for nuclear factors [80]. The maximal increase of serum LBP concentration during the acute phase of trauma or sepsis patients is reached at day 2–3, which fits to the results of animal experiments and to studies employing hepatoma cell lines, where maximum LBP levels were observed 24–48 h after stimulation with cytokines [20, 78]. The acute phase rise in LBP levels appears to be of high predictive value for sepsis as compared to known acute phase markers such as IL-6, pro-calcitonin (pCT) or CRP [own unpubl. results].

By employing an in vitro system utilizing the mouse macrophage cell line RAW 264.7 and by establishing a mouse peritonitis model, we were able to

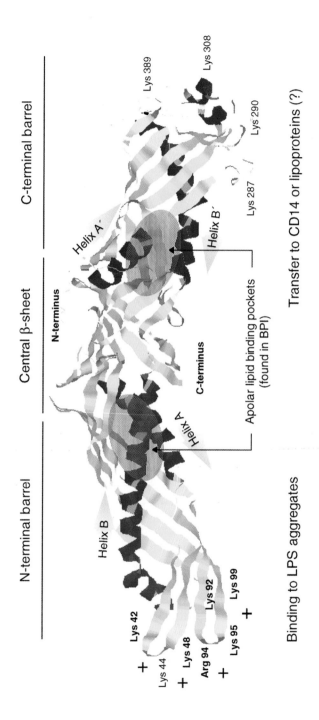

N-terminal barrel   Central β-sheet   C-terminal barrel

Helix A'

Helix A

Helix B

Helix B'

N-terminus

C-terminus

Lys 389

Lys 308

Lys 290

Lys 287

Lys 42

Lys 44

Lys 48

Arg 94

Lys 92

Lys 95

Lys 99

Apolar lipid binding pockets
(found in BPI)

Transfer to CD14 or lipoproteins (?)

Binding to LPS aggregates

1

show recently that acute-phase concentrations of LBP inhibit LPS effects in vitro and protect mice from an otherwise lethal infection [24]. In vitro, LPS fails to induce cytokine production in the murine macrophage cell line when high concentrations of LBP are present. While low concentrations of LBP enhanced TNF-α production of RAW cells, increasing the LBP concentration to acute-phase levels blocked TNF secretion. A parallel i.p. injection of high concentrations of LBP reduced the in vivo cytokine production induced by LPS, and prevented lethality of an LPS injection in mice [24]. These results were supported by recent results obtained by analysis of the LBP knock-out mouse: Jack et al. [46] could show that LBP$^{-/-}$ mice were more susceptible to the lethal consequences of infection with live bacteria than normal mice. LBP here apparently exhibits a protective role as the lack of LBP in the $^{-/-}$ mouse leads to a worsened outcome. LBP in vivo thus may have a dual biological role: initially it may enable the host to rapidly detect and respond to low concentrations of LPS and later, during an acute phase, it may cause suppression of LPS-dependent effects.

Mechanisms of LBP Function

Recognition of Endotoxin Micelles
Bacterial LPS is a cone-shaped amphipathic molecule with a relatively large hydrophobic and a small hydrophilic region [81–83]. The hydrophilic region consists of polysaccharides, composed of the core region and the O-specific chain [84]. The structurally highly conserved lipophilic portion of LPS, lipid A, is the active principle required for endotoxic activities [85] and cytokine induction in host defence cells [86, 87]. Lipid A has been shown to be responsible for the interaction of LPS with recognition proteins, such as LBP [17, 56] or lipoproteins [88]. Moreover, lipid A is essential for bacterial survival [89] and activation of the classical antibody-independent complement cascade [90]. As an amphipathic molecule, LPS forms supramolecular aggregates in aqueous environment, i.e. micelles and cubic structures [82]. LPS is anchored in the aggregate by the lipophilic fatty acid chains of the lipid A portion, whereas the hydrophilic polysaccharide region extends outward from the aggregate

Fig. 1. Computer model of a potential 3-dimensional structure of LBP based on the crystallography of BPI. The coordinates of a computer alignment of BPI and LBP resulting in a likely structure of LBP were obtained via internet from Dr. L. Beamer. The 3-dimensional model was obtained applying the Rasmol software. Marked are several exposed positively charged amino acids at the distal areas of the hypothetical protein structure. Bold-face amino acids represent residues found in both, LBP and BPI. Two shaded areas represent as indicated lipid-binding pockets found in BPI.

surface. Polymeric LPS binds to leukocytes poorly and fails to provoke a response at low concentrations. On addition of high concentrations of LPS micelles in buffered saline, a wide variety of cells are stimulated, suggesting that LPS is able to directly stimulate cellular responses. The addition of plasma, however, has been shown to dramatically accelerate the binding of LPS to cells and to evoke cellular responses at picogram/ml concentrations [91, 92]. The key to understanding LPS recognition in mammals is to be found in the nature of its interaction with binding proteins present in plasma. LBP and CD14 represent the major soluble extracellular proteins that bind to LPS and mediate its biological activity.

LBP Facilitates Monomerization of LPS and Promotes Cellular Activation

It has been shown that monomeric LPS is capable of stimulating cells at low, physiologically relevant concentrations [93]. Since spontaneous diffusion of LPS monomers from aggregates occurs at a very slow rate in plasma [94], effective LPS recognition requires serum factors that mediate monomerization and lead to its dispersal in plasma. LBP binds to LPS with high affinity and has been shown to facilitate the process of LPS monomerization and subsequent presentation to other cellular and humoral binding sites [40, 56]. It catalyzes the transfer of LPS monomers to a binding site on CD14 [18, 94–96] and, when added to serum-free systems, enhances the LPS-mediated stimulation of CD14-positive cells 100- to 1,000-fold [17, 73]. As a member of a family of lipid transfer proteins, LBP furthermore catalyzes the transport of LPS to lipoproteins or phospholipid vesicles either directly [97, 98] or by utilizing sCD14 as a shuttle molecule [58, 99]. As is true for many endogenous mediators [100–106], the LPS receptor CD14 appears in a membrane-bound and at least one soluble form [26, 107, 108]. The ability of LBP to transfer disaggregated LPS to both the membrane-bound and the soluble form of CD14, supports the view that LBP has a central role in mediating LPS responses. LPS monomers, complexed to LBP, potently activate a variety of cell types including both cells that express CD14 on the membrane [22, 109–112], and cells that are mCD14-negative [113–116]. For stimulation of CD14-negative cells, however, soluble CD14 appears to be sufficient for mediatng LPS effects and the requirement for LBP is much less pronounced than with mCD14-positive cells.

LBP catalyzes LPS binding to its cellular receptor consisting of CD14, and most likely members of the Toll protein family: CD14 is anchored to the cell membrane by a glycophosphatidylinositol (GPI) anchor [117] and thus contains no membrane spanning hydrophobic domain [107] capable of directly transmitting a signal into the cell. Moreover, sCD14 has been demonstrated to mediate endotoxin responsiveness in CD14-negative cells, presumably by binding to a coreceptor which mediates transmembrane signaling. In this model, the GPI-anchored mCD14 or the soluble receptor sCD14 bind LPS extracellularly either in the transcellular space or plasma and subsequently activates transmembrane signaling on binding to cell-associated receptor subunits or coreceptors. A mechanism of this type is also found in the receptor-signaling system employed by glial cell line-derived neurotrophic factor (GDNF), a member of the TGF-β ligand family, and in leukemia-inhibiting factor (LIF) signaling. For GDNF, signaling is initiated by the transmembrane receptor tyrosine kinase (RET) that is activated upon binding of GDNF to the GPI-anchored GNDF receptor-α [118]. Most recently, members of the family of Toll-like receptors (TLR) have been suggested to represent a signaling component of the cellular LPS receptor [62, 63]. In the human system, TLR-2 appears to mediate LPS-induced NF-κB activation, while in mice TLR-4 appears to play this role. Two different strains of LPS-hyporesponsive mice have been found to either carry a point mutation, or a complete deletion, respectively, in the TLR-4 gene, strongly supporting the concept of the Toll-receptor family being centrally involved in LPS signaling [64, 65]. As reported for CD14 [119, 120], TLR2 m-RNA was also upregulated following LPS stimulation of mononuclear cells and effective LPS signaling was dependent on LBP and enhanced by CD14 [63].

LBP Catalyzes LPS Transfer

Recent work has described LBP as transferring LPS in a catalytic fashion. It was proposed that a single LBP molecule is able to transport hundreds of LPS molecules to CD14, and that LBP is not consumed in this reaction. Several lines of evidence suggest that LBP is not a stoichiometric partner in the final complexes formed [96]. Yu and Wright [95] demonstrated first order kinetics for this enzymatic transfer and were further able to define catalytic constants for this reaction. It was calculated that the turnover number for LBP is approximately 150 mmol of LPS min^{-1} LBP^{-1} [95]. Two models were proposed for explaining the catalytic reaction mechanism for the transfer of LPS to sCD14: A 'ping pong' or 'binary complex' model suggesting that LBP first interacts with an LPS micelle in a bimolecular reaction and dissociates from the micelle with one molecule LPS bound. This complex binds to sCD14 in a second bimolecular reaction [96]. The second, 'ternary complex' model, proposes a simultaneous interaction of LBP, the LPS micelle, and sCD14 during transfer of LPS monomers to CD14. The authors present several lines of evidence favoring the ternary complex model for transfer of LPS monomers to sCD14 rather than the binary one [95].

LBP-Mediated LPS Detoxification Involves Lipoproteins

In whole plasma, LBP may catalyze movement of LPS to acceptors other than CD14. Incubation of LPS with plasma revealed a LPS-detoxifying capacity of plasma. It has been shown earlier, that upon incubation of LPS with plasma, LPS can no longer cause fever and death in experimental animals [121, 122]. Plasma, furthermore, was able to inhibit LPS signaling in the *Limulus* amebocyte lysate assay [123]. This LPS detoxification seems to occur without chemical modification since LPS retains full biological activity after extraction from plasma with organic solvents [124]. Several studies suggest that lipoprotein particles may reconstitute the LPS-neutralizing activity in plasma: Ulevitch and coworkers were the first to demonstrate the capacity of HDL to bind and neutralize LPS [122, 125], while Van Lenten et al. [126] demonstrated LPS-neutralizing capacity for LDL. For triglyceride-rich lipoproteins (VLDL and chylomicrons) similar capacities were reported [127, 128]. Reconstituted HDL consisting of purified apolipoprotein A-1, phosphatidylcholine, and cholesterol, neutralizes LPS in whole blood more efficiently than LDL, VLDL, or natural HDL [129], while 'intralipid', a triglyceride-rich fat lipid emulsion was shown to enhance LPS stimulatory effects [130].

Reconstituted HDL (R-HDL) added to LPS-containing plasma potently neutralizes LPS, however, reconstituted HDL cannot neutralize LPS by itself in the absence of plasma. Addition of purified LBP to these serum-free systems can restore the LPS-neutralizing capacity of R-HDL. Similarly, as observed with LBP-mediated movement of LPS to CD14, LBP acts catalytically in the transfer of LPS into R-HDL, thus accelerating the rate of LPS detoxification [97]. The transfer of LPS to R-HDL appears to proceed not only directly but also in a two-step reaction in which LPS is first transferred to sCD14, and subsequently shuttled to R-HDL. Moreover, this two-step pathway seems to be strongly favored over the direct transfer, since neutralization of LPS by R-HDL was accelerated more than 30-fold by addition of sCD14. Binding of LPS by lipoprotein particles results in increased LPS clearance [131] and decreased binding of LPS to cells [132]. This, in consequence, leads to diminished production and release of proinflammatory cytokines [133], possibly representing an endogenous clearance mechanism for bacterial LPS. Evidence exists that this mechanism is also operative in vivo, since infusion of lipoproteins [134, 135] or hyperlipoproteinemia in genetically modified mice [136, 137] protects the host against otherwise lethal endotoxemia or severe gram-negative infections. In contrast, hypolipidemic mice are more sensitive to LPS [138]. Levine et al. [136] suggest that the lipid A domain of LPS could be masked by insertion into the phospholipid bilayer of discoidal R-HDL or in the phospholipid monolayer of spherical plasma HDL.

Several lines of evidence support the notion that LBP is physically associated with HDL in vivo: (1) LBP co-purifies with HDL. (2) A passage of plasma over an anti-apoA-1 column removes 99% of plasma LBP. (3) R-HDL requires addition of plasma or LBP to detoxify LPS, whereas apoA-1 containing lipoproteins isolated from plasma by selected affinity immunosorption neutralizes LPS without the need for additional plasma or LBP [60, 97]. (4) A recent study has found nearly all plasma LBP located on a novel lipoprotein particle that additionally contains apoA-1, factor H-related protein 1 and 2 (FHRP-1, FHRP-2), phospholipids and other proteins [60]. The lipid transfer proteins cholesterol ester transfer protein (CETP) and phospholipid transfer protein (PLTP) share sequence homology with LBP [52] and are both found on the surface of HDL [139, 140]. PLTP transfers phospholipids into HDL and CETP transfers cholesterol esters and triglycerides between lipoproteins. Thus, LBP seems to be a member of lipid transferases located on lipoprotein particles that physiologically mediate lipid transfer. During endotoxemia, LBP plays a dual role in both enabling and inhibiting responses to LPS. As LBP catalytically transfers LPS to at least two destinations, CD14 and lipoproteins, this interplay apparently results in a dynamic equilibrium. In such, the kinetics of either reaction, and the molar concentrations of either reactant determine the direction of LPS transfer. A recent study addressed this issue and assessed the kinetics of LPS neutralization by lipoproteins in relation to its binding to and stimulation of human peripheral blood mononuclear cells (PBMC). The authors found that binding of LPS to lipoproteins is much slower than the binding of LPS to PBMC. These data imply the hypothesis that LPS in patients with high LPS concentrations bind to cells and stimulate a potent immune response before neutralization of lipoproteins can occur [141]. However, plasma from patients in the acute phase was shown to possess a reduced ability to stimulate monocytes with LPS [own unpubl. results).

LBP and CD14 Act as Lipid Transfer Proteins

As outlined above, LBP belongs to a family of lipid transfer proteins including phospholipid transfer protein (PLTP), cholesterol ester transfer protein (CETP) and bacterial/permeability increasing protein (BPI) [17, 49, 52, 97]. Recent studies revealed that LBP and CD14 can also function in phospholipid transport. Yu et al. [142] observed transport of phosphatidylinositol (PI), phosphatidylcholine (PC), and a fluorescent derivative of phosphatidylethanolamine (R-PE) from membranes with high concentrations of these phospholipids to reconstituted HDL (rHDL) particles and this transport was mediated by LBP and CD14. These findings define LBP and CD14 as novel phospholipid transfer proteins. As seen in LPS transport, LBP was shown to catalyze the binding of PI and R-PE to sCD14. Upon addition of sCD14, furthermore,

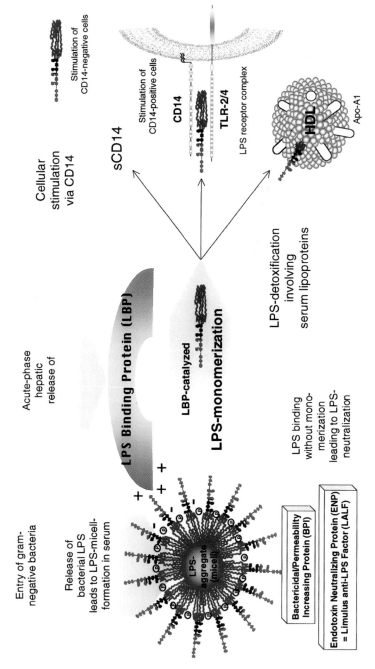

Entry of gram-negative bacteria

Release of bacterial LPS leads to LPS-micell-formation in serum

Acute-phase hepatic release of

LPS Binding Protein (LBP)

LBP-catalyzed LPS-monomerization

LPS-detoxification involving serum lipoproteins

LPS binding without mono-merization leading to LPS-neutralization

Bactericidal/Permeability Increasing Protein (BPI)

Endotoxin Neutralizing Protein (ENP) = Limulus anti-LPS Factor (LALF)

LPS-aggregate (micell)

Cellular stimulation via CD14

sCD14

Stimulation of CD14-negative cells

Stimulation of CD14-positive cells

CD14

TLR-2/4

LPS receptor complex

HDL

Apo-A1

2

the phospholipid transport to rHDL was more efficient. LBP differs from other phosholipid transferases in requiring sCD14 as a soluble carrier protein for optimal phospholipid transfer activity.

CD14 has been suggested to act as a 'pattern recognition receptor' exhibiting binding specificity for foreign molecules that allow CD14 to initiate innate immune responses [33]. This hypothesis was supported by observations showing that CD14 not only binds and transports LPS, but also a broad range of bacterial products including lipoteichoic acid from gram-positive bacteria [143], lipoarabinomannan from *Mycobacterium tuberculosis* [144], amphiphilic membrane molecules and peptidoglycans from *Staphylococcus aureus* [34, 143], mannuronic acid polymers from *Pseudomonas aeruginosa* [145], rhamnose-glucose polymers from *Streptococcus mutans* [146], and chitosans from arthropods [147]. The novel role of LBP and CD14 in phospholipid metabolism, however, demonstrates binding of mammalian phospholipids that have no capacity to initiate innate immune responses. Yu et al. [142] therefore suggest that in mammals LPS is transferred to targets cells by an existing lipid transfer system, namely LBP and sCD14, and is recognized as foreign further downstream, potentially within the cell membrane. Wang et al. [148] recently reported that a series of phosphatidylinosides and phosphoatidylserine exhibit the ability to bind to CD14 and that LBP facilitates binding of these phospholipids to mCD14. Anti-CD14 monoclonal antibodies that inhibit LPS-mCD14 binding also block the binding of phosphatidylinosides to mCD14 [148]. Since phosphatidylinosides can inhibit both, LPS-mCD14 binding and LPS-induced responses in monocytes, it was postulated that endogenous phosphatidylinosides modulate cellular responses to LPS or other mCD14 ligands in vivo. Furthermore, recent observations that human CD14 mediates recognition and phagocytosis of 'self' components like apoptotic cells support the view that CD14 is not solely restricted to the binding of foreign molecules [149]. Recent unpublished observations, furthermore, show that LBP is able to bind to other bacterial ligands, like lipoteichoic acid, and thus may also be involved in 'pattern recognition'. A summary of the function of LBP, LPS monomerization and presentation to either HDL or the LPS receptor complex is shown in figure 2.

Fig. 2. Model for LPS monomerization catalyzed by LBP. After entry of bacteria and cell death, LPS is released and forms multimers (micelles). LPS multimers can be bound by LBP, BPI or ENP. Only LBP apparently has the ability to monomerize LPS and present it to either the cellular LPS receptor consisting of CD14 and a member of the Toll protein family leading to cellular activation, or to HDL particles resulting in detoxification of LPS.

References

1   Shilo M: Non-specific resistance to infection. Ann Rev Microbiol 1959;13:255–270.
2   Schumann R: Recognition of bacterial endotoxin and modulation of the inflammatory response: The LBP/CD14 pathway during the acute phase. Sepsis 1998;2:149–155.
3   Ulevitch RJ, Tobias PS: Recognition of gram-negative bacteria and endotoxin by the innate immune system. Curr Opin Immunol 1999;11:19–22.
4   Medzhitov R, Janeway CA Jr: Innate immunity: Impact on the adaptive immune response. Curr Opin Immunol 1997;9:4–9.
5   Janeway CA Jr, Medzhitov R: Introduction: The role of innate immunity in the adaptive immune response. Semin Immunol 1998;10:349–350.
6   Medzhitov R, Janeway CA Jr: Innate immunity: The virtues of a nonclonal system of recognition. Cell 1997;91:295–298.
7   Kopp EB, Medzhitov R: The toll-receptor family and control of innate immunity. Curr Opin Immunol 1999;11:13–18.
8   Nathan CF: Secretory products of macrophages. J Clin Invest 1987;97:319–329.
9   Johnston RB Jr: Monocytes and macrophages. N Engl J Med 1988;318:747–751.
10 van Furth R: Production and migration of monocytes and kinetics of macrophages; in van Furth R (ed): Mononuclear Phagocytes: Biology of Monocytes and Macrophages. Dordrecht, Kluwer Academic Publishers, 1992, pp 3–12.
11 Knopf H-P, Otto F, Engelhardt R, Freudenberg MA, Galanos C, Herrmann F, Schumann RR: Discordant adaptation of human peritoneal macrophages to stimulation by lipopolysaccharide (LPS) and the synthetic lipid A analogue SDZ MRL 953: Downregulation of TNF-α and IL-6 is paralleled by an upregulation of IL-1β and G-CSF expression. J Immunol 1994;153:287–299.
12 Young LS: Gram negative sepsis; in Mandell G (ed): Principles and Practices of Infectious Diseases. New York, Wiley, 1990, pp 611–635.
13 Glauser MP, Zanetti G, Baumgartner J-D, Cohen J: Septic shock: Pathogenesis. Lancet 1991;338:732–736.
14 Bone RC: The pathogenesis of sepsis. Ann Intern Med 1991;115:457–469.
15 Parillo JE: Pathogenic mechanisms of septic shock. N Engl J Med 1993;328:1471–1477.
16 Tobias P, Soldau K, Ulevitch RJ: Isolation of a lipopolysaccharide-binding acute phase reactant from rabbit serum. J Exp Med 1986;164:777–793.
17 Schumann RR, Leong SR, Flaggs GW, Gray PW, Wright SD, Mathison JC, Tobias PS, Ulevitch RJ: Structure and function of lipopolysaccharide binding protein. Science 1990;249:1429–1431.
18 Wright SD, Ramos RA, Tobias PS, Ulevitch RJ, Mathison JC: CD14, a receptor for complexes of lipopolysaccharide (LPS) and LPS binding protein. Science 1990;249:1431–1433.
19 Bazil V, Strominger JL: Shedding as a mechanism of down-modulation of CD14 on stimulated human monocytes. J Immunol 1991;147:1567–1574.
20 Schumann RR, Kirschning CJ, Unbehaun A, Aberle HP, Knopf HP, Lamping N, Ulevitch RJ, Herrmann F: The lipopolysaccharide-binding protein is a secretory class 1 acute-phase protein whose gene is transcriptionally activated by APRF/STAT/3 and other cytokine-inducible nuclear proteins. Mol Cell Biol 1996;16:3490–3503.
21 Schütt C, Schumann R: Der Endotoxinrezeptor CD14. Immun Infekt 1993;21:36–40.
22 Ziegler-Heitbrock HWL, Ulevitch RJ: CD14: Cell surface receptor and differentiation marker. Immunol Today 1993;14:121–125.
23 Schumann RR, Rietschel ET, Loppnow H: The role of CD14 and lipopolysaccharide-binding protein (LBP) in the activation of different cell types by endotoxin. Med Microbiol Immunol (Berl) 1994;183:279–297.
24 Lamping N, Dettmer R, Schroder NW, Pfeil D, Hallatschek W, Burger R, Schumann RR: LPS-binding protein protects mice from septic shock caused by LPS or gram-negative bacteria. J Clin Invest 1998;101:2065–2071.
25 Schutt C, Schilling T, Kruger C: sCD14 prevents endotoxin inducible oxidative burst response of human monocytes. Allerg Immunol 1991;37:159–164.
26 Frey EA, Miller DS, Jahr TG, Sundan A, Bazil V, Espevik T, Finlay BB, Wright SD: Soluble CD14 participates in the response of cells to lipopolysaccharide. J Exp Med 1992;176:1665–1671.

27   Pugin J, Schurer-Maly CC, Leturcq D, Moriarty A, Ulevitch RJ, Tobias PS: Lipopolysaccharide activation of human endothelial and epithelial cells is mediated by lipopolysaccharide-binding protein and soluble CD14. Proc Natl Acad Sci USA 1993;90:2744–2748.

28   Haziot A, Rong GW, Lin XY, Silver J, Goyert SM: Recombinant soluble CD14 prevents mortality in mice treated with endotoxin (lipopolysaccharide). J Immunol 1995;154:6529–6532.

29   Stelter F, Witt S, Furll B, Jack RS, Hartung T, Schutt C: Different efficacy of soluble CD14 treatment in high- and low-dose LPS models. Eur J Clin Invest 1998;28:205–213.

30   Stelter F, Bernheiden M, Menzel R, Witt S, Jack RS, Grunwald U, Fan X, Schutt C: The molecular basis for therapeutic concepts utilizing CD14. Prog Clin Biol Res 1998;397:301–313.

31   Weidemann B, Brade H, Rietschel ET, Dziarski R, Bazil V, Kusumoto S, Flad H-D, Ulmer AJ: Soluble peptidoglycan-induced monokine production can be blocked by anti-CD14 monoclonal antibodies and by lipid A partial structures. Infect Immun 1994;62:4709–4715.

32   Heumann D, Barras C, Severin A, Glauser MP, Tomasz A: Gram-positive cell walls stimulate synthesis of tumor necrosis factor alpha and interleukin-6 by human monocytes. Infect Immun 1994;62:2715–2721.

33   Pugin J, Heumann ID, Tomasz A, Kravchenko VV, Akamatsu Y, Nishijima M, Glauser MP, Tobias PS, Ulevitch RJ: CD14 is a pattern recognition receptor. Immunity 1994;1:509–516.

34   Gupta D, Kirkland TN, Viriyakosol S, Dziarski RRA: CD14 is a cell activating receptor for bacterial peptidoglycan. J Biol Chem 1996;271:23310–23316.

35   Beamer LJ, Carroll SF, Eisenberg D: Crystal structure of human BPI and two bound phospholipids at 2.4 angstrom resolution. Science 1997;276:1861–1864.

36   Beamer LJ, Carroll SF, Eisenberg D: The BPI/LBP family of proteins: A structural analysis of conserved regions. Protein Sci 1998;7:906–914.

37   Beamer LJ, Carroll SF, Eisenberg D: The three-dimensional structure of human bactericidal/permeability-increasing protein: Implications for understanding protein-lipopolysaccharide interactions. Biochem Pharmacol 1999;57:225–229.

38   Viriyakosol S, Kirkland TN: A region of human CD14 required for lipopolysaccharide binding. J Biol Chem 1995;270:361–368.

39   Juan TS, Hailman E, Kelley MJ, Busse LA, Davy E, Empig CJ, Narhi LO, Wright SD, Lichenstein HS: Identification of a lipopolysaccharide binding domain in CD14 between amino acids 57 and 64. J Biol Chem 1995;270:5219–5224.

40   Lamping N, Hoess A, Yu B, Park TC, Kirschning CJ, Pfeil D, Reuter D, Wright SD, Herrmann F, Schumann RR: Effects of site directed mutagenesis of basic residues (Arg 94, Lys 95, Lys 99) of lipopolysaccharide (LPS) binding protein on binding and transfer of LPS and subsequent immune cell activation. J Immunol 1996;157:4648–4656.

41   Viriyakosol S, Kirkland TN: The N-terminal half of membrane CD14 is a functional cellular lipopolysaccharide receptor. Infect Immun 1996;64:653–656.

42   Stelter F, Bernheiden M, Menzel R, Jack RS, Witt S, Fan X, Pfister M, Schutt C: Mutation of amino acids 39-44 of human CD14 abrogates binding of lipopolysaccharide and *Escherichia coli*. Eur J Biochem 1997;243:100–109.

43   Schumann RR, Lamping N, Hoess A: Interchangeable endotoxin-binding domains in proteins with opposite lipopolysaccharide-dependent activities. J Immunol 1997;159:5599–5605.

44   Kirkland TN, Viriyakosol S: Structure-function analysis of soluble and membrane-bound CD14. Prog Clin Biol Res 1998;397:79–87.

45   Haziot A, Ferrero E, Kontgen F, Hijiya N, Yamamoto S, Silver J, Stewart CL, Goyert SM: Resistance to endotoxin shock and reduced dissemination of gram-negative bacteria in CD14-deficient mice. Immunity 1996;4:407–414.

46   Jack RS, Fan X, Bernheiden M, Rune G, Ehlers M, Weber A, Kirsch G, Mentel R, Furll B, Freudenberg M, Schmitz G, Stelter F, Schutt C: Lipopolysaccharide-binding protein is required to combat a murine gram-negative bacterial infection. Nature 1997;389:742–745.

47   Wurfel MM, Monks BG, Ingalls RR, Dedrick RL, Delude R, Zhou D, Lamping N, Schumann RR, Thieringer R, Fenton MJ, Wright SD, Golenbock D: Targeted deletion of the lipopolysaccharide (LPS)-binding protein gene leads to profound suppression of LPS responses ex vivo, whereas in vivo responses remain intact. J Exp Med 1997;186:2051–2056.

48 Grube BJ, Cochane CG, Ye RD, Green CE, McPhail ME, Ulevitch RJ, Tobias PS: Lipopolysaccharide binding protein expression in primary human hepatocytes and HepG2 hepatoma cells. J Biol Chem 1994;269:8477–8482.

49 Tobias PS, Mathison JC, Ulevitch RJ: A family of lipopolysaccharide binding proteins involved in response to gram-negative sepsis. J Biol Chem 1988;263:13479–13488.

50 Agellon LB, Quinet EM, Gillette TG, Drayna DT, Brown ML, Tall AR: Organization of the human cholesteryl ester transfer protein gene. Biochemistry 1990;29:1372–1376.

51 Au-Young J, Fielding CJ: Synthesis and secretion of wild-type and mutant human plasma cholesteryl ester transfer protein in baculovirus-transfected insect cells: The carboxyl-terminal region is required for both lipoprotein binding and catalysis of transfer. Proc Natl Acad Sci USA 1992;89: 4094–4098.

52 Day JR, Albers JJ, Ce L-D, Gilbert TL, Ching AF, Grant FJ, O'Hara PJ, Marcovina SM, Adolphson JL: Complete cDNA encoding human phospholipid transfer protein from human endothelial cells. J Biol Chem 1994;269:9388–9391.

53 Hubacek JA, Buchler C, Arlaudis C, Schmitz G: The genomic organization of the genes for human lipopolysaccharide binding protein (LBP) and bactericidal permeability increasing protein (BPI) is highly conserved. Biochem Biophys Res Commun 1997;236:427–430.

54 Kirschning CJ, Au-Young J, Lamping N, Reuter D, Pfeil D, Seilhamer JJ, Schumann RR: Similar organization of the lipopolysaccharide-binding protein (LBP) and phospholipid transfer protein (PLTP) genes suggests a common gene family of lipid-binding proteins. Genomics 1997;46:416–425.

55 Elsbach P: The bactericidal/permeability-increasing protein (BPI) in antibacterial host defense. J Leukoc Biol 1998;64:14–18.

56 Tobias PS, Soldau K, Ulevitch RJ: Identification of a lipid A binding site in the acute phase reactant lipopolysaccharide binding protein. J Biol Chem 1989;264:10867–10871.

57 Schumann RR: Function of lipopolysaccharide (LPS)-binding protein (LBP) and CD14, the receptor for LPS/LBP complexes: A short review. Res Immunol 1992;143:11–15.

58 Wurfel MM, Hailman E, Wright SD: Soluble CD14 acts as a shuttle in the neutralization of lipopolysaccharide (LPS) by LPS-binding protein and reconstituted high density lipoprotein. J Exp Med 1995;181:1743–1754.

59 Hailman E, Albers JJ, Wolfbauer G, Tu AY, Wright SD: Neutralization and transfer of lipopolysaccharide by phospholipid transfer protein. J Biol Chem 1996;271:12172–12178.

60 Park CT, Wright SD: Plasma lipopolysaccharide-binding protein is found associated with a particle containing apolipoprotein A-I, phospholipid, and factor H-related proteins. J Biol Chem 1996;271: 18054–18060.

61 Wright SD, Tobias PS, Ulevitch RJ, Ramos RA: Lipopolysaccharide (LPS) binding protein opsonizes LPS-bearing particles for recognition by a novel receptor on macrophages. J Exp Med 1989;170: 1231–1241.

62 Kirschning CJ, Wesche H, Merrill Ayres T, Rothe M: Human toll-like receptor 2 confers responsiveness to bacterial lipopolysaccharide. J Exp Med 1998;188:2091–2097.

63 Yang RB, Mark MR, Gray A, Huang A, Xie MH, Zhang M, Goddard A, Wood WI, Gurney AL, Godowski PJ: Toll-like receptor-2 mediates lipopolysaccharide-induced cellular signalling. Nature 1998;395:284–288.

64 Poltorak A, He X, Smirnova I, Liu MY, Huffel CV, Du X, Birdwell D, Alejos E, Silva MGC, Freudenberg M, Ricciardi-Castagnoli P, Layton B, Beutler B: Defective LPS signaling in C3H/HeJ and C57BL/10ScCr mice: Mutations in tlr4 gene. Science 1998;282:2085–2088.

65 Qureshi ST, Lariviere L, Leveque G, Clermont S, Moore KJ, Gros P, Malo D: Endotoxin-tolerant mice have mutations in Toll-like receptor 4 (Tlr4). J Exp Med 1999;189:615–625.

66 Wright SD: Toll, a new piece in the puzzle of innate immunity. J Exp Med 1999;189:605–609.

67 Ulevitch RJ: Endotoxin opens the Tollgates to innate immunity. Nat Med 1999;5:144–145.

68 Lamping N, Hoess A, Yu B, Park TC, Wright SD, Kirschning CJ, Pfeil D, Schumann RR: Identification of the lipopolysaccharide (LPS) binding site of LPS binding protein (LBP) by site-directed mutagenesis: Evidence for a similar LPS recognition mechanism in different LPS binding proteins; in Faist E (ed): The Immune Consequences of Trauma, Shock and Sepsis, Mechanisms and Therapeutic Approaches. Bologna, Monduzzi Editore, 1997, pp 15–19.

69   Hoess A, Watson S, Siber GR, Liddington R: Crystal structure of an endotoxin-neutralizing protein from horseshoe crab, limulus anti-LPS factor, at 1.5 A resolution. EMBO J 1993;12:3351–3356.

70   Gray PW, Flaggs G, Leong SR, Gumina RJ, Weiss J, Ooi CE, Elsbach P: Cloning of the cDNA of a human neutrophil bactericidal protein: Structural and functional correlations. J Biol Chem 1989;264:9505–9509.

71   Weiss J, Elsbach P, Shu C, Castillo J, Grinna L, Horwitz A, Theofan G: Human bactericidal/permeability-increasing protein and a recombinant NH_2-terminal fragment cause killing of serum-resistant gram-negative bacteria in whole blood and inhibit tumor necrosis factor release induced by the bacteria. J Clin Invest 1992;90:1122–1130.

72   Elsbach P, Weiss J: Prospects for use of recombinant BPI in the treatment of gram-negative bacterial infections. Infect Agents Dis 1995;4:102–109.

73   Mathison JC, Tobias PS, Wolfson E, Ulevitch RJ: Plasma lipopolysaccharide (LPS)-binding protein: A key component in macrophage recognition of gram-negative LPS. J Immunol 1992;149:200–206.

74   Mathison JC, Virca GD, Wolfson E, Tobias PS, Glaser K, Ulevitch RJ: Adaptation to bacterial lipopolysaccharide controls lipopolysaccharide-induced tumor necrosis factor production in rabbit macrophages. J Clin Invest 1990;85:1108–1117.

75   Grunwald U, Fan XL, Jack RS, Workalemahu G, Kallies A, Stelter F, Schütt C: Monocytes can phagocytose gram negative bacteria by a CD14 dependent mechanism. J Immunol 1996;157: 4119–4125.

76   Vosbeck K, Tobias P, Mueller H, Allen RA, Arfors KE, Ulevitch RJ, Sklar LA: Priming of polymorphonuclear granulocytes by lipopolysaccharides and its complexes with lipopolysaccharide binding protein and high density lipoprotein. J Leukocyte Biol 1990;47:97–104.

77   Kirschning CJ, Unbehaun A, Fiedler G, Hallatschek W, Lamping N, Pfeil D, Schumann RR: The transcriptional activation pattern of lipopolysaccharide binding protein (LBP) involving transcription factors AP-1 and C/EBP beta. Immunobiology 1997;198:124–135.

78   Kirschning C, Unbehaun A, Lamping N, Pfeil D, Herrmann F, Schumann RR: Control of transcriptional activation of the lipopolysaccharide binding protein (LBP) gene by proinflammatory cytokines [published erratum appears in Cytokines Cell Mol Ther 1997;3:137]. Cytokines Cell Mol Ther 1997; 3:59–62.

79   Fey GH, Gauldie J: The acute phase response of the liver in inflammation; in Popper H, Schaffner F (eds): Progress in Liver Disease. Philadelphia, Saunders, 1990, pp 89–116.

80   Baumann H, Gauldie J: The acute phase response. Immunol Today 1994;15:74–80.

81   Raetz CRH: Biochemistry of endotoxins. Annu Rev Biochem 1990;59:129–170.

82   Seydel U, Labischinski H, Kastowsky M, Brandenburg K: Phase behaviour, supramolecular structure, and molecular conformation of lipopolysaccharide. Immunobiology 1993;187:191–211.

83   Brandenburg K, Seydel URA, Schromm AB, Loppnow H, Koch M, Rietschel ET: Conformation of lipid a, the endotoxic center of bacterial lipopolysaccharide. J Endotoxin Res 1996;3:173–178.

84   Rietschel ET, Brade H: Bacterial endotoxins. Sci Am 1992;267:54–63.

85   Galanos C, Lüderitz O, Rietschel ET, Westphal O, Brade H, Brade L, Freudenberg M, Schade U, Imoto M, Yoshimura H, Kusumoto S, Shiba T: Synthetic and natural E. coli free lipid A express identical endotoxic activities. Eur J Biochem 1985;148:1–5.

86   Loppnow H, Brade L, Brade H, Rietschel ET, Kusumoto S, Shiba T, Flad H-D: Induction of human interleukin-1 by bacterial and synthetic lipid A. Eur J Immunol 1986;16:1263–1267.

87   Loppnow H, Brade H, Dürrbaum I, Dinarello CA, Kusumoto S, Rietschel ET, Flad HD: IL1 induction capacity of defined lipopolysaccharide partial structures. J Immunol 1989;142:3229–3238.

88   Freudenberg MA, Bog-Hansen TC, Back U, Galanos C: Interaction of lipopolysaccharides with plasma high density lipoproteins in rats. Infect Immun 1980;28:373–380.

89   Osborn MJ: Biosynthesis and assembly of lipopolysaccharide of the outer membrane; in Inouye M (ed): Bacterial Outer Membranes, Biogenesis and Functions. New York, Wiley, 1979, pp 15–33.

90   Ihara I, Harada Y, Ihara S, Kawakami M: A new complement-dependent bactericidal factor found in nonimmune mouse sera. J Immunol 1982;128:1256–1260.

91   Ulevitch RJ, Tobias PS: Receptor-dependent mechanisms of cell stimulation by bacterial endotoxin. Annu Rev Immunol 1995;13:437–457.

92   Wright SD: CD14 and innate recognition of bacteria. J Immunol 1995;155:6–8.
93   Takayama K, Mitchell DH, Din ZZ, Mukerjee P, Li C, Coleman DL: Monomeric Re lipopolysaccha-
     ride from *Escherichia coli* is more active than the aggregated form in the *Limulus* amebocyte lysate
     assay and in inducing Egr-1 messenger RNA in murine peritoneal macrophages. J Biol Chem 1994;
     269:2241–2244.
94   Hailman E, Lichtenstein HS, Wurfel MM, Miller DS, Johnson DA, Kelley M, Busse LA, Zukowski
     MM, Wright SD: Lipopolysaccharide (LPS)-binding protein accelerates the binding of LPS to
     CD14. J Exp Med 1994;179:269–277.
95   Yu B, Wright SD: Catalytic properties of lipopolysaccharide (LPS) binding protein (LBP):Transfer
     of LPS to soluble CD14. J Biol Chem 1996;271:4100–4105.
96   Tobias PS, Soldau K, Gegner JA, Mintz D, Ulevitch RJ: Lipopolysaccharide binding protein-
     mediated complexation of lipopolysaccharide with soluble CD14. J Biol Chem 1995;270:10482–
     10488.
97   Wurfel MM, Kunitake ST, Lichtenstein H, Kane JP, Wright SD: Lipopolysaccharide (LPS)-binding
     protein is carried on lipoproteins and acts as a cofactor in the neutralization of LPS. J Exp Med
     1994;180:1025–1035.
98   Vasselon T, Pironkova R, Detmers PA: Sensitive responses of leukocytes to lipopolysaccharide
     require a protein distinct from CD14 at the cell surface. J Immunol 1997;159:4498–4505.
99   Wurfel MM, Wright SD: Lipopolysaccharide binding protein and soluble CD14 transfer lipopoly-
     saccharide to phospholipid bilayers: Preferential interaction with particular classes of lipid. J Im-
     munol 1997;158:3925–3934.
100 Heaney ML, Golde DW: Soluble hormone receptors. Blood 1993;82:1945–1948.
101 Heaney ML, Golde DW: Soluble cytokine receptors. Blood 1996;87:847–857.
102 Rose-John S, Ehlers M, Grotzinger J, Mullberg J: The soluble interleukin-6 receptor. Ann NY Acad
     Sci 1995;762:207–220, discussion 220–201.
103 Rose-John S, Heinrich PC: Soluble receptors for cytokines and growth factors: Generation and
     biological function. Biochem J 1994;300:281–290.
104 Diez-Ruiz A, Tilz GP, Zangerle R, Baier-Bitterlich G, Wachter H, Fuchs D: Soluble receptors for
     tumour necrosis factor in clinical laboratory diagnosis. Eur J Haematol 1995;54:1–8.
105 Arend WP: Inhibiting the effects of cytokines in human diseases. Adv Intern Med 1995;40:365–394.
106 Dinarello CA: Biologic basis for interleukin-1 in disease. Blood 1996;87:2095–2147.
107 Haziot A, Chan S, Ferrero E, Low MG, Silber R, Goyert SM: The monocyte differentiation antigen,
     CD14, is anchored to the cell membrane by a phosphatidyl-inositol linkage. J Immunol 1988;141:
     547–552.
108 Durieux JJ, Vita N, Popescu O, Guette F, Calzada WJ, Munker R, Schmidt RE, Lupker J, Ferrara
     P, Ziegler HH, et al: The two soluble forms of the lipopolysaccharide receptor, CD14: Characteriza-
     tion and release by normal human monocytes. Eur J Immunol 1994;24:2006–2012.
109 Ball E, Graziano RF, Shen L, Fanger MW: Monoclonal antibodies to novel myeloid antigens reveal
     human granulocyte heterogeneity. Proc Natl Acad Sci USA 1982;79:5374–5378.
110 Morabito F, Prasthofer EF, Dunlap NE, Grossi CE, Tilden AB: Expression of myelomonocytic anti-
     gens on chronic lymphocytic leukemia B cells correlates with their ability to produce interleukin 1.
     Blood 1987;70:1750–1757.
111 Calvo F, Martin PM, Jabrane N, Cremoux RD, Magdelenat H: Human breast cancer cells share
     antigens with the myeloid monocyte lineage. Br J Cancer 1987;56:15–19.
112 Ziegler-Heitbrock HWL, Appl B, Kafferlein E, Loffler T, Jahn HH, Gutensohn W, Nores JR,
     McCullough K, Passlick B, Labeta MO, et al: The antibody MY4 recognizes CD14 on porcine
     monocytes and macrophages. Scand J Immunol 1994;40:509–514.
113 Erwin AL, Mandrell RE, Munford RS: Enzymatically deacylated *Neisseria* lipopolysaccharide
     (LPS) inhibits murine splenocyte mitogenesis induced by LPS. Infect Immun 1991;59:1881–1887.
114 Pohlman TH, Munford RS, Harlan JM: Deacylated lipopolysaccharide inhibits neutrophil adherence
     to endothelium induced by lipopolysaccharide in vitro. J Exp Med 1987;165:1393–1402.
115 Riedo FX, Munford RS, Campbell WB, Reisch JS, Chien KR, Gerard RD: Deacylated lipopoly-
     saccharide inhibits plasminogen activator inhibitor-1, prostacyclin, and prostaglandin E2 production
     by lipopolysaccharide but not by tumor necrosis factor. J Immunol 1990;144:3506–3512.

116   Schönbeck U, Flad HD, Rietschel ET, Brandt E, Loppnow H: S-form LPS induces leukocyte adhesion to human vascular endothelial cells as potent as IL1: Lipid A precursor Ia antagonizes induction of adhesion by LPS. J End Res 1994;1:4–13.

117   Simmons DL, Tan S, Tenen DG, Nicholson-Weller A, Seed B: Monocyte antigen CD14 is a phospholipid anchored membrane protein. Blood 1989;73:284–289.

118   Robertson K, Mason I: The GNDF-RET signaling pathway. Trends Genet 1997;13:1–3.

119   Matsuura K, Ishida T, Setoguchi M, Higuchi Y, Akizuki S, Yamamoto S: Upregulation of mouse Cd14 expression in kupffer cells by lipopolysaccharide. J Exp Med 1994;179:1671–1676.

120   Marchant A, Duchow J, Delville JP, Goldman M: Lipopolysaccharide induces up-regulation of CD14 molecule on monocytes in human whole blood. Eur J Immunol 1992;22:1663–1665.

121   Rall DP G, JR, Kelley MG: Reduction of febrile response to bacterial polysaccharide following incubation with serum. Am J Physiol 1957;188:559.

122   Ulevitch RJ, Johnston AR: The modification of biophysical and endotoxic properties of bacterial lipopolysaccharides by serum. J Clin Invest 1978;62:1313–1324.

123   Johnson KJ, Ward P, Goralnick BS, Osborn MJ: Isolation from human serum of an inactivator of bacterial lipopolysaccharide. Am J Pathol 1977;88:559–574.

124   Rudbach JR, Johnson AG: Restoration of endotoxin activity following alteration by plasma. Nature 1964;202:811.

125   Ulevitch RJ, Johnston AR, Weinstein DB: New function for high density lipoproteins: Their participation in intravascular reactions of bacterial lipopolysaccharides (LPS). J Clin Invest 1979;64:1516–1524.

126   Van Lenten BJ, Fogelman AM, Haberland ME, Edwards P: The role of lipoproteins and receptor-mediated endocytosis in the transport of bacterial lipopolysaccharide. Proc Natl Acad Sci USA 1986;83:2704–2708.

127   Harris HW, Grunfeld C, Feingold KR, Rapp JH: Human very low density lipoproteins and chylomicrons can protect against endotoxin-induced death in mice. J Clin Invest 1990;86:696–702.

128   Harris HW, Grunfeld C, Feingold KR, Read TE, Kane JP, Jones AL, Eichbaum EB, Bland GF, Rapp JH: Chylomicrons alter the fate of endotoxin, decreasing tumor necrosis factor release and preventing death. J Clin Invest 1993;91:1028–1034.

129   Parker TS, Levine DM, Chang JC, Laxer J, Coffin CC, Rubin AL: Reconstituted high-density lipoprotein neutralizes gram-negative bacterial lipopolysaccharides in human whole blood. Infect Immun 1995;63:253–258.

130   van der Poll T, Braxton CC, Coyle SM, Boermeester MA, Wang JC, Jansen PM, Montegut WJ, Calvano SE, Hack CE, Lowry SF: Effect of hypertriglyceridemia on endotoxin responsiveness in humans. Infect Immun 1995;63:3396–3400.

131   Read TE, Grunfeld C, Kumwenda ZL, Calhoun MC, Kane JP, Feingold KR, Rapp JH: Triglyceride-rich lipoproteins prevent septic death in rats. J Exp Med 1995;182:267–272.

132   Munford RS, Andersen JM, Dietschy JM: Sites of tissue binding and uptake in vivo of bacterial lipopolysaccharide-high density lipoprotein complexes: Studies in the rat and squirrel monkey. J Clin Invest 1981;68:1503–1513.

133   Flegel WA, Wolpl A, Mannel DN, Northoff H: Inhibition of endotoxin-induced activation of human monocytes by human lipoproteins. Infect Immun 1989;57:2237–2245.

134   Hubsch AP, Powell FS, Lerch PG, Doran JE: A reconstituted, apolipoprotein A-I containing lipoprotein reduces tumor necrosis factor release and attenuates shock in endotoxemic rabbits. Circ Shock 1993;40:14–23.

135   Pajkrt D, Doran JE, Koster F, Lerch PG, Arnet B, van der Poll T, ten Cate JW, van Deventer SJ: Antiinflammatory effects of reconstituted high-density lipoprotein during human endotoxemia. J Exp Med 1996;184:1601–1608.

136   Levine DM, Parker TS, Donnelly TM, Walsh A, Rubin AL: In vivo protection against endotoxin by plasma high density lipoprotein. Proc Natl Acad Sci USA 1993;90:12040–12044.

137   Netea MG, Demacker PN, Kullberg BJ, Boerman OC, Verschueren I, Stalenhoef AF, van der Meer JW: Low-density lipoprotein receptor-deficient mice are protected against lethal endotoxemia and severe gram-negative infections. J Clin Invest 1996;97:1366–1372.

Lipopolysaccharide-Binding Protein   59

138 Feingold KR, Funk JL, Moser AH, Shigenaga JK, Rapp JH, Grunfeld C: Role for circulating lipoproteins in protection from endotoxin toxicity. Infect Immun 1995;63:2041–2046.

139 Pattnaik NM, Zilversmit DB: Interaction of cholesteryl ester exchange protein with human plasma lipoproteins and phospholipid vesicles. J Biol Chem 1979;254:2782–2786.

140 Tall AR, Forester LR, Bongiovanni GL: Facilitation of phosphatidylcholine transfer into high density lipoproteins by an apolipoprotein in the density 1.20–1.26 g/ml fraction of plasma. J Lipid Res 1983;24:277–289.

141 Netea MG, Demacker PN, Kullberg BJ, Jacobs LE, Verver-Jansen TJ, Boerman OC, Stalenhoef AF, Van der Meer JW: Bacterial lipopolysaccharide binds and stimulates cytokine-producing cells before neutralization by endogenous lipoproteins can occur. Cytokine 1998;10:766–772.

142 Yu B, Hailman E, Wright SD: Lipopolysaccharide binding protein and soluble CD14 catalyze exchange of phospholipids. J Clin Invest 1997;99:315–324.

143 Kusunoki T, Hailman E, Juan T, Lichenstein HS, Wright SD: Molecules from *staphylococcus aureus* that bind CD14 and stimulate innate immune responses. J Exp Med 1995;182:1673–1682.

144 Zhang Y, Doerfler M, Lee TC, Guillemin B, Rom WN: Mechanisms of stimulation of interleukin-1 beta and tumor necrosis factor-alpha by *Mycobacterium tuberculosis* components. J Clin Invest 1993; 91:2076–2083.

145 Espevik T, Otterlei M, Skjak BG, Ryan L, Wright SD, Sundan A: The involvement of CD14 in stimulation of cytokine production by uronic acid polymers. Eur J Immunol 1993;23:255–261.

146 Soell M, Lett E, Holveck F, Scholler M, Wachsmann D, Klein JP: Activation of human monocytes by streptococcal rhamnose glucose polymers is mediated by CD14 antigen, and mannan binding protein inhibits TNF-alpha release. J Immunol 1995;154:851–860.

147 Otterlei M, Varum KM, Ryan L, Espevik T: Characterization of binding and TNF-alpha-inducing ability of chitosans on monocytes: The involvement of CD14. Vaccine 1994;12:825–832.

148 Wang PY, Kitchens RL, Munford RS: Phosphatidylinositides bind to plasma membrane CD14 and can prevent monocyte activation by bacterial lipopolysaccharide. J Biol Chem 1998;273:24309–24313.

149 Devitt A, Moffatt OD, Raykundalia C, Capra JD, Simmons DL, Gregory CD: Human CD14 mediates recognition and phagocytosis of apoptotic cells. Nature 1998;392:505–509.

Ralf R. Schumann, MD, Institut für Mikrobiologie und Hygiene, Charité University Medical Center, Humboldt University, Dorotheenstrasse 96, D–10117 Berlin (Germany)
Tel. +49 30 2093 4747, Fax +49 30 2093 4794, E-Mail ralf.schumann@charite.de

Jack RS (ed): CD14 in the Inflammatory Response.
Chem Immunol. Basel, Karger, 2000, vol 74, pp 61–82

....

Role of CD14 in Cellular Recognition of Bacterial Lipopolysaccharides

Richard L. Kitchens

Department of Internal Medicine, The University of Texas Southwestern Medical
Center, Dallas, Tex., USA

When bacteria invade host tissues, they immediately trigger sensitive innate immune mechanisms that play important roles in both first lines of defense and adaptive immune responses [1]. Unlike mechanisms of adaptive immunity which depend upon lymphocyte antigen receptor rearrangements, clonal selection, and acquired memory, innate immunity depends upon molecular mechanisms of microbial recognition that are 'hard-wired' or programmed into cells and upon memory that has resulted from selective pressure exerted by pathogens over evolutionary time. Because innate mechanisms cannot generate the breadth of diversity seen in adaptive responses, Medzhitov and Janeway [1] proposed that the innate immune system must recognize molecular patterns of conserved products of microbial metabolism rather than particular molecules and that these patterns must be absolutely distinct from self antigens. A variety of 'pattern recognition receptors' have been proposed to recognize microbial structures which include bacterial lipopolysaccharides, lipoproteins, lipotechoic acids, lipoarabinomannan, cell walls, DNA containing unmethylated C_pG motifs, yeast mannans, and viral double-stranded RNA.

Lipopolysaccharide (LPS or endotoxin), an essential component of gram-negative bacteria, can be viewed as a class of related molecules that are constructed according to a common architectural pattern [2]. LPS resides in the outer membranes of these bacteria and is among the most potent bacterial agonists known. Sensitive host mechanisms can detect and respond to picomolar concentrations of LPS in the cellular environment by recognizing the conserved lipid A portion of the LPS molecule. The resulting cellular signals are then amplified to generate an array of mediators that mobilize the immune system to wall off and destroy the invading bacteria. Although LPS is not inherently toxic to cells, its ability to overstimulate immune effector functions can be toxic to the host, and frequently results in coagulation disorders, multiple organ failure, shock, and death.

Research over the past decade has established the importance of CD14 as an LPS receptor that is largely responsible for the sensitivity of host cells to this bacterial lipid [3–6]. Evidence that CD14 can also bind and potentiate cell responses to several other conserved bacterial structures led to the concept that CD14 is a pattern recognition receptor [7]. Recent observations that CD14 can also bind certain nonstimulatory host structures raise the possibility that CD14 has a physiologic role distinct from host defense. The results of research into the binding and functional interactions between CD14 and its ligands (principally LPS) will be reviewed here, and the role of CD14 in microbial recognition will be discussed.

Functional Consequences of LPS-CD14 Interactions

CD14 was first described as a monocyte/macrophage differentiation antigen that is attached to the cell surface by a glycosylphosphatidylinositol (GPI) anchor [8]. Constitutive expression of membrane-bound CD14 (mCD14) is found predominantly on monocytes and macrophages, and to a lesser extent on neutrophils. Human monocytes express high levels of CD14 ($\sim 10^5$ receptors/cell) whereas expression on neutrophils is much lower ($\sim 3 \times 10^3$ receptors/cell) [9]. CD14 expression on tissue macrophages can vary depending upon the source [10]. Fearns et al. [11] showed that CD14 gene expression is much more widespread than had been previously suspected. In situ hybridization revealed that nonmyeloid cells in a variety of tissues express CD14 mRNA either constitutively or inducibly in response to inflammatory stimuli. Whether mCD14 is expressed on the surfaces of these cells has not yet been confirmed. Soluble CD14 (sCD14) occurs in normal human plasma at concentrations of 2–6 µg/ml and is found in two or more anchorless forms [12, 13]. Another important protein in LPS recognition is LPS-binding protein (LBP), a 60-kD plasma glycoprotein that has extensive sequence homology with bactericidal permeability increasing protein (BPI), cholesterol ester transfer protein (CETP), and phospholipid transfer protein (PTLP). LBP is normally found in plasma in low (µg/ml) concentrations, but during the acute-phase response, plasma levels increase many fold [4, 14].

mCD14 Confers Sensitive Responses to LPS
The first evidence of a functional role for CD14 was the discovery that it could bind and enhance the responsiveness of cells to LPS [3]. LBP, a protein

that had been previously identified as a plasma LPS binding protein that recognizes lipid A [4], was also shown to be required for sensitive LPS responses in macrophages and whole blood [14]. Numerous in vitro studies indicate that the cooperative effects of CD14 and LBP greatly increase the sensitivity of cells to low concentrations of LPS. Certain anti-CD14 or anti-LBP antibodies that inhibit LPS binding also inhibit cell activation, and high concentrations of LPS are generally required to overcome the suppressive effects of the antibodies [3, 14]. In vivo studies also show that antibodies to either LBP [15] or CD14 [16] can suppress LPS responses and rescue animals from the lethal effects of LPS.

Stable expression of CD14 cDNA confers sensitivity to LPS in certain cells that do not otherwise express this receptor. When the murine pre-B (70Z/3) cells were transfected with CD14, their sensitivity to LPS was increased up to 10,000-fold [17]. CD14 transfection conferred LPS responsiveness in Chinese hamster ovary (CHO) cells [18] and human fibrosarcoma cells [19], which do not normally respond to LPS. Conversely, LPS hyporesponsiveness is exhibited in murine macrophage mutants that are defective in CD14 expression [20] and in monocytes from paroxysmal nocturnal hemaglobinuria (PNH) patients, which are defective in their ability to make GPI-anchor modifications on proteins [21].

Targeted deletion of the CD14 gene in mice confirmed the importance of CD14 in LPS responsiveness [6]. Monocytes and macrophages derived from these animals are profoundly hyporesponsive to LPS. Modest amounts of TNFα and only trivial amounts of IL-1β and IL-6 were produced in response to high (1 and 10 μg/ml) concentrations of LPS, which are 4–5 logs above the threshold concentrations for LPS responses in cells from control animals. The animals were also highly resistant to LPS in two assays of LPS-induced lethality. These and other studies also confirmed, however, that CD14 is not absolutely required for sensing the presence of LPS [6, 22]. As had been previously suspected, LPS is recognized by CD14-independent mechanisms that can induce cytokine responses and lethality if high enough LPS concentrations are present. Surprisingly, acute-phase proteins are induced normally by LPS in the absence of CD14 [23]. Obvious candidates for these mechanisms are the Toll-like receptors (discussed below), which are thought to cooperate with CD14 in LPS recognition [24–26]. Minimal responses to LPS in the absence of CD14 may also require other binding receptors such as β_2-integrins [27].

Targeted deletion of the LBP gene in mice is also accompanied by impaired sensitivity to LPS. Jack et al. [28] showed that LBP-deficient mice (sensitized with *D*-galactosamine) were resistant to LPS-induced lethality and failed to mount a serum TNFα response to the LPS challenge. As expected, LBP-deficient serum failed to promote binding of FITC-LPS to murine CD14 expressed on CHO cells. When the animals were challenged by infection with gram-negative bacteria (*Salmonella typhimurium*), LBP was required for the survival of the animals. Inflammatory cells from the peritoneal cavities of infected mice failed to respond to the whole bacteria added ex vivo in the presence of LBP-deficient serum, whereas control serum containing LBP conferred responsiveness to the bacteria. Whether LBP is required for in vivo LPS responses has been questioned by Wurfel et al. [29]. They found that the sensitivity to injected LPS was the same for induction of serum TNFα in both LBP-deficient and control animals, whereas production of TNFα in whole blood ex vivo required LBP.

sCD14 Can Augment or Inhibit Cell Responses to LPS
sCD14 was first described as an LPS inhibitor by Schütt et al. [30] according to its ability to inhibit LPS-induced oxidative burst in monocytes. When added to whole human blood at high concentrations ($\sim$70 µg/ml), sCD14 inhibits LPS-induced TNFα production [31]. It has also been reported to rescue mice from LPS-induced mortality [32]. High concentrations of sCD14 may act as an LPS acceptor 'sink' that inhibits LPS interactions with mCD14 [33, 34]. Another anti-inflammatory mechanism of sCD14 involves its ability to accelerate the transfer of LPS to HDL particles, which sequester and inactivate LPS [35]. LBP and PLTP also participate in the sequestration of LPS by transferring LPS from micelles [36, 37] or bacterial membrane fragments [Vesy et al., unpubl. data] to plasma lipoproteins.
sCD14 can potentiate LPS responses in some cells that do not express mCD14 when sCD14 is added at low (serum equivalent) concentrations [33, 38, 39]. LPS-sCD14 complexes activate endothelial cells to express adhesion molecules [33, 38, 39] and cytokines [39]. LPS-sCD14 complexes also activate some (but not all) epithelial cells [39], human astrocytoma cells (U373) [38], and smooth muscle cells [40].

Modes of LPS Binding to CD14

CD14 can bind LPS in at least three different physical settings (fig. 1): (a) LPS in the membranes of intact bacteria; (b) LPS in aggregates or micelles, and (c) LPS monomers.

CD14 Recognizes LBP-Opsonized Gram-Negative Bacteria (fig. 1, A)
Wright et al. [41] first showed that LBP could opsonize LPS-bearing particles (LPS-coated erythrocytes or intact *Salmonella*) and enable them to attach to a novel receptor on macrophages. In experiments using LPS-coated

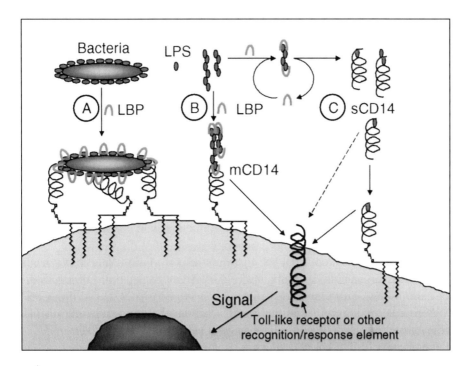

Fig. 1. CD14 binds LPS in at least 3 physical settings. CD14 binds LBP-opsonized gram-negative bacteria and augments phagocytosis (A). CD14 binds LPS aggregates in LPS-LBP complexes (B) and LPS monomers (C), and facilitates LPS or LPS-CD14 interactions with one or more recognition/response elements that transduce signals into the cell.

erythrocytes, they identified the receptor as CD14 [3]. Binding studies by Jack et al. [42] showed that either mCD14 or sCD14 bound to gram-negative bacteria that had been opsonized by serum, and Grunwald et al. [43] demonstrated the ability of monocytes to phagocytose *E. coli* by an LBP and mCD14-dependent mechanism that did not depend upon opsonization of the bacteria with IgG or complement. The roles of LBP and mCD14 in phagocytosis of *E. coli* were studied in detail by Schiff et al. [44] who showed a positive correlation between the level of mCD14 expression on cells and their phagocytic activity toward LBP-opsonized *E. coli*.

These data suggest that the high levels of mCD14 frequently found on monocytes and macrophages can serve to facilitate phagocytosis of bacteria. Such high levels of mCD14 expression do not seem to be required, however, for sensitive responses to free LPS. For example, neutrophils have sensitive CD14-dependent mechanisms of LPS recognition [45] even though these cells express very low levels of mCD14 [9]. Likewise, when we expressed low levels of CD14 in CHO cells using a minimal promoter, the CD14 conferred sensitivity to low concentrations of LPS in NF-κB activation assays, although mCD14 was almost undetectable by flow cytometry [Wang et al., unpubl.].

CD14 Recognizes Complexes Formed by
LPS Aggregates and LBP (fig. 1, B)

Numerous studies have demonstrated LBP-dependent binding of radioactive or fluorescent LPS to mCD14 [17, 46–49], and Tobias et al. [50] showed that LPS, when derivatized with a labeled photoactivatable cross-linker (^{125}I-ASD-LPS), cross-links to mCD14 in an LBP-dependent manner. In the absence of LBP, only trivial amounts of LPS bind to cells nonspecifically. Binding analyses show that [^{3}H]LPS binds to mCD14 with high apparent affinity ($K_d = 8$–49 nM), and at maximal binding levels, the molar ratio of LPS:mCD14 ranged from 1 to 20 under various conditions [48, 49, 51]. Gegner et al. [52] showed that LPS, LBP, and mCD14 form ternary complexes on the cell surface, which can be internalized by the cell. When increasing concentrations of LPS are added to cells at a constant LPS: LBP ratio, LPS: LBP complexes appear to self-associate on the cell surface, resulting in the binding of 1,000 or more molecules of LPS per molecule of mCD14.

Most of the LPS and LBP that bind to mCD14 do not appear to be involved in signaling. An anti-LBP antibody (18G4) can block approximately 90% of the LPS binding to mCD14 without inhibiting signaling. Conversely, a CD14 antibody (18E12) can block LPS signaling without reducing LPS binding to mCD14 [52].

CD14 Recognizes LPS Monomers that Can Be Rapidly Formed by the
Action of LBP (fig. 1, C)

In addition to its opsonizing and complex-forming activities, LBP can also disaggregate LPS and transfer LPS monomers to CD14 [53, 54]. When a molar excess of sCD14 is mixed with LPS, low concentrations of LBP can act catalytically to transfer LPS to sCD14, producing stable complexes that contain 1 or 2 molecules of LPS per molecule of sCD14. Under these conditions, no stable ternary LPS-LBP-sCD14 complexes are formed, and LBP is reused to transfer other molecules of LPS.

However, LBP is not absolutely required for the formation of LPS-sCD14 complexes, although these complexes are formed very slowly in the absence of LBP [53]. Once they are formed, the LPS in LPS-sCD14 complexes can be rapidly transferred to empty sCD14 molecules without a requirement for LBP [55]. mCD14 serves as an acceptor for LPS in LPS-sCD14 complexes and mediates activation of macrophages and neutrophils by these complexes [55] without a requirement for LBP. When [^{3}H]LPS-sCD14 complexes are presented to cells that express mCD14, the [³H]LPS binds rapidly to the cells and appears to occupy mCD14 molecules in an approximately 1:1 molar ratio under saturating conditions, whereas in the absence of sCD14, little or no cell binding of LPS occurs [51]. Thus, for most cell responses, the functional importance of LBP resides in its ability to dramatically increase the rate of binding of LPS to CD14 [56] and thereby increase the rate of availability of the LPS to the cell. Interestingly, however, the expression of at least one LPS-inducible gene (IP-10) requires LBP [57].

Cells that do not express mCD14 may possess an as yet unidentified receptor for sCD14 or LPS-sCD14 complexes. Little [39] or no [51] demonstrable [³H]LPS binding occurs in CD14-negative cells. Tapping and Tobias [58] showed that LBP and LPS were required to demonstrate binding of ³²P-sCD14 to CD14-negative epithelial cells. The importance of sCD14 binding to the cells was unclear in these experiments, however, since LPS and sCD14 were sufficient to activate the cells without LBP. It is possible that cell activation does not require binding of sCD14 to a cellular receptor; sCD14 may instead transfer LPS to a cellular protein [59] or signal transducing receptor (e.g. a Toll-like receptor [24]) that may be expressed at low levels on cell surfaces.

CD14 Structure-Function Relationships

The N-terminal half (152 amino acids) of sCD14 [60] or mCD14 [61] is sufficient for binding LPS and promoting cell responses. Several hydrophilic regions of mCD14 between amino acids 7 and 65 have been shown to be important for LPS binding and cell activation [62, 63]. Protease protection studies and deletion mutagenesis of sCD14 identified an important LPS binding domain at amino acids 57–64 [64, 65]. Alanine substitution mutations in this region of mCD14, however, had no effect on LPS binding or function [63]. Another study by Juan et al. [34] identified a sCD14 domain near its N-terminus that is essential for LPS signaling but not for LPS binding. These results are in contrast to those of Viriyakosol and Kirkland [62] who showed that mCD14 with a deletion in the same N-terminal region failed to bind LPS. These conflicting data raise the possibilities that (a) the same domains of sCD14 and mCD14 can have different functions, and (b) different methods of mutagenesis may produce different results [63]. Another level of complexity of CD14 structure-function was revealed by Shapiro et al. [66] who showed that a negative to positive charge reversal mutation in a single residue of sCD14 abolished binding of *P. gingivalis* LPS but not *E. coli* LPS. Interactions of LPS with CD14 domains are complex, and precise knowledge of how CD14 and lipid A make contact will have to await the solution of the crystal structure of CD14.

The role of the GPI anchor in CD14 function is still uncertain. Experiments in which the GPI anchor of CD14 was replaced with various transmembrane domains showed that the GPI anchor is not essential for CD14-mediated LPS signaling or phagocytosis [44, 67–69].

Relationship of LPS Signaling to LPS Internalization and Intracellular Traffic

It is clear that much of the LPS that binds to CD14 becomes internalized [51, 52, 70–72], and there is general agreement that LPS internalization by phagocytes serves to sequester LPS and limit its interactions with other cells. For a detailed discussion of LPS internalization and degradation, the reader is referred to a recent review [73].

The relationship between LPS internalization and signaling has been a matter of controversy. Some investigators argue that the bulk of the LPS that binds to CD14 is destined for sequestration and disposal, whereas a very small fraction of the LPS is involved in productive signaling interactions on the cell surface [52, 58]. Others have suggested that endocytic movement of LPS plays a role in signaling [59, 72, 74–76]. Thieblemont and Wright [75] reported that macrophages from LPS hyporesponsive C3H/HeJ mice exhibit a defect in intracellular traffic of LPS; specifically, internalized LPS fails to coalesce normally in a perinuclear location. Subsequently, these investigators reported that LPS from *Rhodobacter sphaeroides* (RsLPS), a nonstimulatory LPS antagonist, exhibits defective LPS traffic in normally responsive cells and inhibits intracellular traffic of a stimulatory form of LPS [76]. Although these studies show an interesting relationship between LPS signaling and traffic, they do not provide strong evidence to support the hypothesis that intracellular movement of LPS is required for signaling. The alternative interpretation, that LPS signaling influences intracellular traffic, was not ruled out. They also reported that exposure of cells to chlorpromazine and certain other cationic amphipaths converted that LPS antagonist (RsLPS) into an agonist and restored normal intracellular traffic of RsLPS. To explain these data, the investigators hypothesized that chlorpromazine alters the packing geometry of RsLPS molecules that are intercalated into the plasma membrane. They also proposed that biophysical interactions of LPS with membrane lipids dictate both the nature of LPS intracellular transport and is ability to stimulate cellular responses. A similar biophysical hypothesis of LPS signaling was proposed by Schromm et al. [77] who showed that lipid A charge and conformation influence the ability of LPS to intercalate into phospholipid vesicles in the presence of LBP and to stimulate IL-6 in blood cells. These interesting new hypotheses are similar to a longstanding hypothesis that fell into disfavor with many scientists when attention turned to the investigation of putative lipid A receptors on the cell surface; the hypothesis was that insertion of LPS into the lipid bilayer of the plasma membrane perturbs membrane structure in a way that sets up a signal response [78]. It is conceivable that interactions with specific receptors and membrane lipids work together to generate specificity in cellular recognition of LPS.

The insertion of LPS into plasma membranes is difficult to measure with confidence. Studies of CD14-dependent LPS internalization from our lab showed that most surface-exposed LPS remains protein bound [51]. Very little binding of [³H]LPS from LPS-sCD14 complexes can be measured in CD14-negative cells [39, 51], whereas plasma lipoproteins or phospholipid vesicles rapidly bind LPS under the same conditions [35; 79; Kitchens and Munford, unpubl.]. Although cell membranes are largely resistant to LPS insertion, only a small fraction of the cell-associated LPS may be required to produce signal responses, as noted above. It is also possible that membrane insertion of LPS may occur in an intracellular compartment. Some creative new approaches will, therefore, be needed to test the hypothesis of Wright and coworkers [76].

Studies of CD14-dependent LPS internalization in our lab showed that the first step of this process (movement of LPS across the plasma membrane) is not influenced by the ability of cells to respond to LPS [51]. Macrophages from C3H/HeN and C3H/HeJ mice internalized LPS identically, and a lipid A partial structure that inhibited cell responses to LPS, did not influence LPS internalization. (These observations do not conflict with the studies of Thieblemont, that show alterations in intracellular traffic of internalized LPS [75, 76].) LPS aggregation, however, accelerated its internalization but had no effect on its signal potency [51]. We also found that monocytes can internalize LPS by at least two CD14-dependent pathways. The bulk of LPS appears to enter noncoated tubular invaginations of the plasma membrane, whereas endocytosis by clathrin-coated pits appears to be a minor pathway [69]. Entry of LPS into these noncoated caveolae-like invaginations may be related to the observation that LPS binds to mCD14 in low-density lipid-enriched domains of the plasma membrane that have properties similar to those of caveolae [80]. A mCD14 fusion protein containing the transmembrane and coated-pit targeting domains of the low-density lipoprotein receptor (CD14-LDLR) diverted both aggregated and monomeric forms of LPS to the coated pit pathway but did not influence its signal potency. A large fraction of the LPS that bound to CD14-LDLR, however, entered noncoated tubular membrane invaginations before entering coated pits [69]. CD14-LDLR also failed to exclude LPS from caveolae-like membrane fractions [Wang et al. unpubl.]. Another study also provided evidence that LPS can enter monocytes by macropinocytosis [81]. More research will be required to determine the functional significance of the different endocytic mechanisms and intracellular pathways taken by LPS.

Lipid A Recognition/Response Elements beyond CD14

Insights from Inhibitory Mechanisms of LPS Partial Structures
Structural changes in lipid A that decrease its bioactivity can also convert this molecule into a potent and specific LPS antagonist. The secondary fatty acyl chains of lipid A can be selectively removed by the action of the leukocyte enzyme, acyloxyacyl hydrolase [82] to generate enzymatically deacylated LPS (dLPS). Pohlman et al. [83] showed that dLPS inhibited the ability of LPS to induce adhesion of neutrophils to endothelial cells in an LPS-specific manner. Similar LPS-specific inhibitory properties of dLPS and isolated (lipid IVA) or chemically defined (compound 406, LA-14-PP) tetraacyl lipid A analogs were described in other response assays using a variety of human cells [47, 84]. The competitive nature of the interactions between LPS and these analogs suggests the involvement of one or more specific lipid A receptors.

LPS partial structures were initially thought to antagonize LPS by blocking its binding to mCD14 [85]. Research from our group, however, revealed that the inhibitory mechanism is more complex. In fact, at least three mechanisms can be involved [49]. The most interesting of these was revealed by the ability of low (1–3 ng/ml) concentrations of dLPS or LA-14-PP to inhibit cell responses in an LPS-specific manner while having no inhibitory effect on the binding of [^{3}H]LPS to mCD14 on THP-1 cells [47, 49]. An analysis of [^{3}H]dLPS binding to mCD14 showed that it was saturable (even in physiologic medium), and the apparent dissociation constant (K_d or concentration required to occupy half of the available binding sites) was approximately 100 ng/ml. At 1–3 ng/ml, however, the dLPS concentration is too far below its binding K_d to inhibit LPS binding to CD14. Rossignol and colleagues [86] also found that low concentrations (10–100 nM) of the synthetic lipid A analog, E5531, could completely block TNFα release while failing to measurably inhibit CD14-dependent LPS binding in normal human monocytes. These data suggest that dLPS and lipid A partial structures can somehow inhibit LPS interactions with a response element that is distal to CD14 in the LPS recognition pathway. The events that occur after LPS binds to CD14 are still unclear, however, and a more detailed knowledge of the molecular mechanisms of LPS recognition will be required to determine precisely how LPS partial structures inhibit this process.

Other inhibitory mechanisms of LPS partial structures include competition with LPS for binding to CD14 and/or LBP. Investigators from the Borstel group [85, 87] showed that high concentrations of tetraacyl lipid A analogs inhibited CD14-dependent binding of labeled LPS to human monocytes. High concentrations of dLPS and other lipid A analogs also inhibit LPS binding to CD14 and LBP [49, 88, 89]. Since analog concentrations required to inhibit LPS binding are considerably higher than those required to inhibit signaling, inhibition of LPS binding to mCD14 may be a less important mechanism for inhibiting functional interactions of LPS with cells.

Basis of Ligand-Specific Recognition of Lipid A Analogs

The ability of animals (or their cells) to recognize a particular type of lipid A partial structure depends, in part, upon the species of the animal. For example, lipid A analogs that lack secondary fatty acyl chains fail to stimulate human cells, whereas they are quite stimulatory to mouse or hamster cells in most in vitro assays. LPS or lipid A derived from *Rhodobacter* (*Rhodopseudomonas*) *spheroides* (RSLA or RsDPLA) [90], *Rhodobacter capsulatus* [91], or analogs based on its structure (E5531) [89] fail to stimulate human or mouse cells, but can stimulate hamster cells normally [19]. A recently produced analog of the same class, B1287, fails to stimulate cells from hamsters as well as from mice and humans [92].

To determine whether species-specific differences in CD14 structure were responsible for recognition of structural differences in lipid A, DeLude et al. [19] measured the stimulatory and antagonistic activities of structurally diverse forms of lipid A mediated by mouse or human CD14 expressed in cells from different species. Human (HT-1080) cells were transfected with either human or mouse CD14, mouse (70Z/3) cells were transfected with human CD14, and hamster (CHO) cells were transfected with human or mouse CD14. The results showed that the ability of the cells to recognize structural differences in lipid A was imparted by the genetic background of the cell and not by structural differences in CD14. The authors hypothesized that the target of the LPS antagonists is a lipid A recognition protein that transduces the LPS signal after interacting with CD14-bound LPS. Ingalls et al. [27] also found that overexpression of human β_2-integrin (CD11/CD18) cDNA in CHO cells conferred responsiveness to LPS. In these cells also, the LPS mimetic (lipid IVA, RSLA) and antagonistic (B1287) properties of the lipid A analogs were determined by the cell background and not by the integrin receptors [92].

Candidates from the Toll-Like Receptor Family

Although CD14 is required to confer sensitivity in LPS recognition, other cellular molecules are required to complete the process of cellular recognition and response. Major lines of evidence for the recognition/response elements discussed above include (a) the fact that GPI-anchored and soluble CD14 do not have transmembrane signaling domains; (b) LPS analogs can bind to CD14 without stimulating the cell; (c) LPS-specific antagonists can inhibit LPS responses without inhibiting binding to CD14, and (d) CD14 cannot discriminate between structurally diverse forms of lipid A.

Several recent studies have generated much interest in the potential roles of an emerging family of Toll-like receptors (TLR) in LPS recognition. Yang et al. [24] and Kirschning et al. [25] showed that overexpression of human TLR2 in LPS nonresponsive HEK-293 cells conferred the ability of LPS to activate NF-κB. The sensitivity of TLR2-expressing cells to LPS was increased either by serum, LBP, sCD14, or by co-expression of mCD14. Using a positional cloning approach, Poltorak et al. [26] showed that LPS-hyporesponsive C3H/HeJ and C57BL/10ScCr mice have mutations in the TLR4 gene found at the LPS^d locus on chromosome 4. At the time of writing of this review, it is unclear whether any of the Toll proteins are LPS receptors. Yang et al. [24] reported that [^{3}H]LPS bound slowly to a soluble IgG-TLR2 fusion protein, whereas others failed to demonstrate LPS binding to 293 cells that overexpressed the protein [25]. It is clear, however, that C3H/HeJ macrophages can bind LPS normally [93]. This is most likely due to CD14 expression as macrophages from both C3H/HeJ and C57BL/10ScCr mice express CD14 mRNA and protein normally [94], but they fail to respond to LPS presumably due to a mutation in (HeJ) or lack of expression of (ScCr) TLR4 [26].

Non-LPS Ligands of CD14

The search for non-LPS CD14 ligands has revealed that CD14 can bind a variety of diverse structures derived from both microbes and animal hosts (table 1). The first non-LPS ligands were described by Espevik et al. [95] as β1-4-linked polyuronic acids from bacteria, fungi, and oxidized cellulose.

Pugin et al. [7] showed that mCD14 expression increased the dose sensitivity of cells to gram-positive bacterial cell walls or mycobacterial lipoarabinomannan (LAM) by 30- to 300-fold. The dose sensitivity to cell walls was the same in macrophages from C3H/FeJ (LPS responsive) and C3H/HeJ (LPS nonresponsive) mice, and LAM was also reported to stimulate HeJ cells. Interestingly, antibodies to murine CD14 inhibited responses to cell walls or LAM in both FeJ and HeJ cells, indicating that CD14 promotes interactions with response elements other than the LPS gene product, TLR4 [26]. Medvedev et al. [96] also showed that macrophages from C3H/OuJ (LPS^n) and C3H/HeJ (LPS^d) mice responded with equal sensitivity to cell walls from

Table 1. Non-LPS CD14 ligands

Ligand	Source	Ref. No.
Microbial structures		
Cell walls	*Bacillus subtilis*	7
	Staphylococcus aureus	
	group B *Streptococci*	96
Soluble peptidoglycan	*Staphylococcus aureus*	98, 99, 101
Rhamnose glucose polymers	*Streptococcus mutans*	111
β1-4-linked polyuronic acids	*Pseudomonas syringae*	95
	Pseudomonas aeruginosa	
	Ascophyllum nodosum	
Lipoarabinomannan	*Mycobacterium tuberculosis*	7, 97, 112
Lipotechoic acids (LTA)	*Staphylococcus aureus*	102, 103
	Streptococcus pyogenes	
LTA-like molecules	*Staphylococcus aureus*	104, 105
Lipoproteins	*Borrelia burgdorferi*	106, 107
Whole microorganisms		
	Mycobacterium tuberculosis	113
WI-1 antigen	*Blastomyces dermatitidis*	114
	Cryptococcus neoformans	115
Host structures		
Phospholipids		79, 108, 109
Apoptotic cells		110
Cytokine-stimulated endothelial cells		116
Interleukin-2		117

group B streptococci by a mechanism that involves both CD14 and β_2-integrin receptors. LBP also enhanced cellular responses to LAM [97].

Soluble peptidoglycan, the basic structural component of bacterial cell walls, has been shown to bind and activate cells through CD14 [98–101]. Studies with synthetic muramyl dipeptides (MDP) have shown that these peptidoglycan subunits are the minimal structures required for binding to CD14 [100, 101]. Dziarski et al. [101] found that soluble peptidoglycan binds to sCD14 and mCD14, whereas MDP monomers do not. MDP binds to CD14 only when the MDP is polymerized, aggregated, or immobilized on a solid support. Although CD14 appears to have a broad ligand-binding specificity for certain microbial polymers, its inability to bind a variety of other polymers such as dextrans and heparin indicates that it is not a general receptor for polysaccharides and other biological polymers [101].

Lipotechoic acids (LTA) from gram-positive bacteria have also been reported to stimulate cell responses (e.g. NO and IL-12) by a mechanism that is enhanced by CD14 [102, 103]. However, Kusunoki and coworkers [104, 105] isolated a heterogeneous set of amphipathic molecules from *Staphylococcus aureus* membranes and showed that these molecules bind to sCD14 and stimulate cells with much greater potency than crude LTA preparations. These investigators also claimed that highly purified LTA binds sCD14 but has no stimulatory activity.

Membrane lipoproteins isolated from the spirochete, *Borrelia burgdorferi*, and synthetic lipopeptides based on the structures of lipoproteins from *Treponema pallidum* and *B. burgdorferi* stimulate NF-κB activation and cytokine production in phagocytes [106] and endothelial cells [107] by mechanisms that are enhanced by mCD14 and sCD14, respectively. Sellati et al. [106] showed that peritoneal macrophages from CD14-deficient mice were almost completely insensitive to native OspA, and they were much less sensitive to synthetic lipopeptides than cells from control mice. Previous research by the same group also showed that these lipoprotein and lipopeptide preparations stimulated macrophages from both C3H/HeN (LPS responsive) and C3H/HeJ (LPS non-responsive) mice. Taken together, these studies suggest that CD14 can promote interactions of these spirochetal agonists with cellular response elements that are distinct from TLR4.

Recent studies show that the ligand-binding specificity of CD14 is not restricted to 'non-self' microbial structures. Wright and coworkers have shown that both sCD14 and LBP can act as transfer proteins for certain host-derived phospholipids [79, 108], and Wang et al. [109] also showed that phosphatidylinositides and phosphatidylserine (PS) bind to mCD14 by an LBP-dependent mechanism. mCD14 has recently been implicated as a mechanism for binding and phagocytosis of apoptotic cells [110]. The ability of mCD14 to bind PS, which is externalized during apoptosis, may account for its ability to bind these host cells. Recognition of host molecules by CD14 and LBP raises the possibility that they have physiologic roles apart from microbial recognition.

Insights into the Nature of Microbial Pattern Recognition Provided by CD14

The literature reveals several important aspects of CD14 interactions with LPS and other microbial molecules that contribute to the understanding of microbial pattern recognition. (a) CD14 recognizes molecular patterns in several conserved microbial structures (e.g. lipopolysaccharides, lipoproteins, lipo-

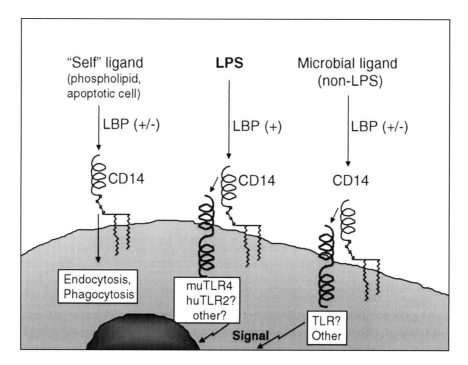

"Self" ligand
(phospholipid,
apoptotic cell)

LPS

Microbial ligand
(non-LPS)

LBP (+/-)

LBP (+)

LBP (+/-)

CD14

CD14

CD14

Endocytosis,
Phagocytosis

muTLR4
huTLR2?
other?

TLR?
Other

Signal

Fig. 2. Components of cellular recognition of LPS and other CD14 ligands. LBP facilitates binding of LPS and certain other ligands to CD14. CD14 facilitates interactions of microbial ligands with multiple recognition/response elements, whereas CD14 binds and facilitates endocytic uptake (e.g. phagocytosis) of host-derived ligands.

arbinomannan, LTA, and cell walls). (b) CD14 has a profound impact on the ability of the animal host to respond to LPS from gram-negative bacteria, and by virtue of its broad ligand-binding specificity, CD14 also plays a role in recognition of other microbial structures. (c) The ligand-binding specificity of CD14 does not, by itself, confer specificity in cellular recognition of microbial molecules. In other words, CD14 can bind stimulatory and nonstimulatory structures, whereas cell responses are mediated by other putative receptors of more refined specificity. (d) CD14 appears to work in concert with multiple cellular response elements. The latter conclusion is based on the known ability of CD14 to augment cell responses to non-LPS ligands in macrophages from C3H/HeJ (LPS-nonresponsive) mice. Since defects in TLR4 are evidently responsible for defective LPS responses in these mice, non-LPS ligands must be recognized by other response elements (e.g. TLR family members or other receptors).

Cellular recognition of LPS can, thus, be viewed as a process that involves at least three pattern recognition receptors that work in concert – LBP, CD14, and presumably a Toll-like receptor or other recognition/response element (fig. 2). CD14 also participates in the recognition of non-LPS microbial structures in conjunction with recognition/response elements that are distinct from those in the LPS pathway. Although these and other mechanisms of innate immune recognition must discriminate perfectly between self and nonself [1], this discrimination does not occur at the level of CD14. In this kind of multi-component system, it should not be essential for each and every pattern recognition receptor involved to possess this degree of specificity. For example, self structures that are recognized by CD14 and LBP are nonstimulatory and evidently are not detected by distal recognition/response elements. On the other hand, if CD14 is absent, then the more specific recognition/response elements fail to recognize LPS for lack of sufficient sensitivity.

Therefore, it appears that the CD14 system is not analogous to many cytokine receptor complexes in which the binding chain confers specificity and common chains (shared among several cytokine receptors) generate signals. Rather, in some respects, CD14 is more analogous to molecules of the major histocompatability complex (MHC). Both CD14 and MHC molecules have broad but restricted ligand binding specificities that include self as well as nonself molecules. This property allows these receptors to sample their environment and present their ligands to other receptors of greater specificity (TLRs or other recognition/response elements, or T-cell receptors, respectively), thereby completing the process of nonself recognition and stimulating effector functions. The manner of presentation of LPS to TLRs or other recognition/response elements (i.e. whether LPS remains bound to CD14 [4] or is transferred from CD14 [5]) remains to be determined.

A more complete understanding of LPS interactions with LBP, CD14, and other pattern recognition receptors should provide an interesting paradigm for innate microbial recognition. An important challenge for the next decade will be to determine how CD14 and its various ligands interact with TLRs and other microbial pattern recognition molecules.

Acknowledgments

I thank Drs. Robert Munford, Tim Sellati, and Deborah Bouis for critical reading of the manuscript. This work was supported by NIH grant AI 18188.

References

1   Medzhitov R, Janeway CA: Innate immunity: Impact on the adaptive immune response. Curr Opin Immunol 1997;9:4–9.
2   Rietschel ET, Brade H, Holst O, Brade L, Müller-Loennies S, Mamat U, Zähringer U, Beckmann F, Seydel U, Brandenburg K, Ulmer AJ, Mattern T, Heine H, Schletter J, Loppnow H, Schönbeck U, Flad H-D, Hauschildt S, Schade UF, Di Padova F, Kusumoto S, Schumann RR: Bacterial endotoxin: Chemical composition, biological recognition, host response, and immunological detoxification; in Rietschel ET, Wagner H (eds): Current Topics in Microbiology and Immunology, vol 216. Berlin, Springer, 1996, pp 39–81.
3   Wright SD, Ramos RA, Tobias PS, Ulevitch RJ, Mathison JC: CD14, a receptor for complexes of lipopolysaccharide (LPS) and LPS binding protein. Science 1990;249:1431–1433.
4   Ulevitch RJ, Tobias PS: Receptor-dependent mechanisms of cell stimulation by bacterial endotoxin. Annu Rev Immunol 1995;13:437–457.
5   Wright SD: CD14 and innate recognition of bacteria. J Immunol 1995;155:6–8.
6   Haziot A, Ferrero E, Köntgen F, Hijiya N, Yamamoto S, Silver J, Stewart CL, Goyert SM: Resistance to endotoxin shock and reduced dissemination of gram-negative bacteria in CD14-deficient mice. Immunity 1997;4:407–414.
7   Pugin J, Heumann D, Tomasz A, Kravchenko VV, Akamatsu Y, Nishijima M, Glauser MP, Tobias PS, Ulevitch RJ: CD14 is a pattern recognition receptor. Immunity 1994;i:509–516.
8   Haziot A, Chen S, Ferrero E, Low MG, Silber R, Goyert SM: The monocyte differentiation antigen, CD14, is anchored to the cell membrane by a phosphatidylinositol linkage. J Immunol 1988;141:547–552.
9   Antal-Szalmas P, Van Strijp JAG, Weersink AJL, Verhoef J, Van Kessel KPM: Quantitation of surface CD14 on human monocytes and neutrophils. J Leukocyte Biol 1997;61:721–728.
10 Ziegler-Heitbrock HWL, Ulevitch RJ: CD14: Cell surface receptor and differentiation marker. Immunol Today 1993;14:121–125.
11 Fearns C, Kravchenko VV, Ulevitch RJ, Loskutoff DJ:: Murine CD14 gene expression in vivo: Extramyeloid synthesis and regulation by lipopolysaccharide. J Exp Med 1995;181:857–866.
12 Bazil V, Baudy M, Hilgert I, Stefanova I, Low MG, Zbrozek J, Horejsi V: Structural relationship between the soluble and membrane-bound forms of human monocyte surface glycoprotein CD14. Mol Immunol 1989;26:657–662.
13 Grunwald U, Kruger C, Westerman J, Lukowsky A, Ehlers M, Schütt C: An enzyme-linked immunosorbent assay for the quantification of solubilized CD14 in biological fluids. J Immunol Methods 1992;155:225–232.
14 Schumann RR, Leong SR, Flaggs GW, Gray PW, Wright SD, Mathison JC, Tobias PS, Ulevitch RJ: Structure and function of lipopolysaccharide-binding protein. Science 1990;249:1429–1431.
15 Gallay P, Heumann D, Le Roy D, Barras C, Glauser M-P: Mode of action of anti-lipopolysaccharide-binding protein antibodies for prevention of endotoxemic shock in mice. Proc Natl Acad Sci USA 1994;91:7922–7926.
16 Leturcq DJ, Moriarty AM, Talbott G, Winn RK, Martin TR, Ulevitch RJ: Antibodies against CD14 protect primates from endotoxin-induced shock. J Clin Invest 1996;98:1533–1538.
17 Lee J-D, Kato K, Tobias PS, Kirkland TN, Ulevitch RJ: Transfection of CD14 into 70Z/3 cells dramatically enhances the sensitivity to complexes of lipopolysaccharide (LPS) and LPS binding protein. J Exp Med 1992;175:1697–1705.
18 Golenbock DT, Liu Y, Millham FH, Freeman MW, Zoeller RA: Surface expression of human CD14 in Chinese hamster ovary fibroblasts imparts macrophage-like responsiveness to bacterial endotoxin. J Biol Chem 1993;268:22055–22059.
19 Delude RL, Savedra R Jr, Zhao HL, Thieringer R, Yamamoto S, Fenton MJ, Golenbock DT: CD14 enhances cellular responses to endotoxin without imparting ligand-specific recognition. Proc Natl Acad Sci USA 1995;92:9288–9292.
20 Nishijima M, Hara-Kuge S, Takasuka N, Akagawa K, Setouchi M, Matsuura K, Yamamoto S, Akamatsu Y: Identification of a biochemical lesion, and characteristic response to lipopolysaccharide (LPS) of a cultured macrophage-like cell mutant with defective LPS-binding. J Biochem (Tokyo) 1994;116:1082–1087.

21   Duchow J, Marchant A, Crusiaux A, Husson C, Alonso-Vega C, De Groote D, Neve P, Goldman M: Impaired phagocyte responses to lipopolysaccharide in paroxysmal nocturnal hemoglobinuria. Infect Immun 1993;61:4280–4285.

22   Perera PY, Vogel SN, Detore GR, Haziot A, Goyert SM: CD14-dependent and CD14-independent signaling pathways in murine macrophages from normal and CD14 knockout mice stimulated with lipopolysaccharide or taxol. J Immunol 1997;158:4422–4429.

23   Haziot A, Lin XY, Zhang F, Goyert SM: Cutting edge: The induction of acute phase proteins by lipopolysaccharide uses a novel pathway that is CD14-independent. J Immunol 1998;160:2570–2572.

24   Yang R-B, Mark MR, Gray A, Huang A, Xie MH, Zhang M, Goddard A, Wood WI, Gurney AL, Godowski PJ: Toll-like receptor-2 mediates lipopolysaccharide-induced cellular signaling. Nature 1998;395:284–288.

25   Kirschning CJ, Wesche H, Ayres TM, Rothe M: Human Toll-like reeptor 2 confers responsiveness to bacterial lipopolysaccharide. J Exp Med 1998;188:2091–2097.

26   Poltorak A, He X, Smirnova I, Liu M-YvHC, Du X, Birdwell D, Alejos E, Silva M, Galanos C, Freudenberg M, Ricciardi-Castagnoli P, Layton B, Beutler B: Defective LPS signaling in C3H/HeJ and C57BL/10ScCr mice: Mutations in *Tlr4* gene. Science 1998;282:2085–2088.

27   Ingalls RR, Arnaout MA, Golenbock DT: Outside-in signaling by lipopolysaccharide through a tailless integrin. J Immunol 1997;159:433–438.

28   Jack RS, Fan X, Bernhelden M, Rune G, Ehlers M, Weber A, Kirsch G, Mentel R, Fürll B, Freudenberg M, Schmitz G, Stelter F, Schütt C: Lipopolysaccharide-binding protein is required to combat a murine gram-negative bacterial infection. Nature 1997;389:742–744.

29   Wurfel MM, Monks BG, Ingalls RR, Dedrick RL, Delude R, Zhou D, Lamping N, Schumann RR, Thieringer R, Fenton MJ, Wright SD, Golenbock D: Targeted deletion of the lipopolysaccharide (LPS)-binding protein leads to profound suppression of LPS responses ex vivo, whereas in vivo responses remain intact. J Exp Med 1997;186:2051–2056.

30   Schütt C, Schilling T, Kruger C: sCD14 prevents endotoxin inducible oxidative burst response of human monocytes. Allerg Immunol (Leipz) 1991;37:159–164.

31   Haziot A, Rong G-W, Bazil V, Silver J, Goyert SM: Recombinant soluble CD14 inhibits LPS-induced tumor necrosis factor-α production by cells in whole blood. J Immunol 1994;152:5868–5876.

32   Haziot A, Rong GW, Lin XY, Silver J, Goyert SM: Recombinant soluble CD14 prevents mortality in mice treated with endotoxin (lipopolysaccharide). J Immunol 1995;154:6529–6532.

33   Haziot A, Rong GW, Silver J, Goyert SM: Recombinant soluble CD14 mediates the activation of endothelial cells by lipopolysaccharide. J Immunol 1993;151:1500–1507.

34   Juan TSC, Hailman E, Kelley MJ, Wright SD, Lichenstein HS: Identification of a domain in soluble CD14 essential for lipopolysaccharide (LPS) signaling but not LPS binding. J Biol Chem 1995;270:17237–17242.

35   Wurfel MM, Hailman E, Wright SD: Soluble CD14 acts as a shuttle in the neutralization of lipopolysaccharide (LPS) by LPS-binding protein and reconstituted high density lipoprotein. J Exp Med 1995;181:1743–1754.

36   Wurfel MM, Kunitake ST, Lichenstein H, Kane JP, Wright SD: Lipopolysaccharide (LPS)-binding protein is carried on lipoproteins and acts as a cofactor in the neutralization of LPS. J Exp Med 1994;180:1025–1035.

37   Hailman E, Albers JJ, Wolfbauer G, Tu AY, Wright SD: Neutralization and transfer of lipopoly-saccharide by phospholipid transfer protein. J Biol Chem 1996;271:12172–12178.

38   Frey EA, Miller DS, Jahr TG, Sundan A, Bazil V, Espevik T, Finlay BB, Wright SD: Soluble CD14 participates in the response of cells to lipopolysaccharide. J Exp Med 1992;176:1665–1671.

39   Pugin J, Schürer-Maly C-C, Leturcq D, Moriarty A, Ulevitch RJ, Tobias PS: Lipopolysaccharide activation of human endothelial and epithelial cells is mediated by lipopolysaccharide-binding protein and soluble CD14. Proc Natl Acad Sci USA 1993;90:2744–2748.

40   Loppnow H, Stelter F, Schönbeck U, Schlüter C, Ernst M, Schütt C, Flad H-D: Endotoxin activates human vascular smooth muscle cells despite lack of expression of CD14 mRNA or endogenous membrane CD14. Infect Immun 1995;63:1020–1026.

41   Wright SD, Tobias PS, Ulevitch RJ, Ramos RA: Lipopolysaccharide (LPS) binding protein opsonizes LPS-bearing particles for recognition by a novel receptor on macrophages. J Exp Med 1989;170: 1231–1241.

42   Jack RS, Grunwald U, Stelter F, Workalemahu G, Schütt C: Both membrane-bound and soluble forms of CD14 bind to gram-negative bacteria. Eur J Immunol 1995;25:1436–1441.

43   Grunwald U, Fan XL, Jack RS, Workalemahu G, Kallies A, Stelter F, Schütt C: Monocytes can phagocytose gram-negative bacteria by a CD14-dependent mechanism. J Immunol 1996;157: 4119–4125.

44   Schiff DE, Kline L, Soldau K, Lee JD, Pugin J, Tobias PS, Ulevitch RJ: Phagocytosis of gram-negative bacteria by a unique CD14-dependent mechanism. J Leukocyte Biol 1997;62:786–794.

45   Wright SD, Ramos RA, Hermanowski-Vosatka A, Rockwell P, Detmers PA: Activation of the adhesive capacity of CR3 on neutrophils by endotoxin: Dependence on lipopolysaccharide binding protein and CD14. J Exp Med 1991;173:1281–1286.

46   Heumann D, Gallay P, Barras C, Zaech P, Ulevitch RJ, Tobias PS, Glauser M-P, Baumgartner JD: Control of lipopolysaccharide (LPS) binding and LPS-induced tumor necrosis factor secretion in human peripheral blood monocytes. J Immunol 1992;148:3505–3512.

47   Kitchens RL, Ulevitch RJ, Munford RS: Lipopolysaccharide (LPS) partial structures inhibit responses to LPS in a human macrophage cell line without inhibiting LPS uptake by a CD14-mediated pathway. J Exp Med 1992;1760:485–494.

48   Kirkland TN, Finley F, Leturcq D, Moriarty A, Lee J-D, Ulevitch RJ, Tobias PS: Analysis of lipopolysaccharide binding by CD14. J Biol Chem 1993;268:24818–24823.

49   Kitchens RL, Munford RS: Enzymatically deacylated lipopolysaccharide (LPS) can antagonize LPS at multiple sites in the LPS recognition pathway. J Biol Chem 1995;270:9904–9910.

50   Tobias PS, Soldau K, Kline L, Le J-D, Kato K, Martin TP, Ulevitch RJ: Cross-linking of lipopolysaccharide (LPS) to CD14 on THP-1 cells mediated by LPS-binding protein. J Immunol 1993;150: 3011–3021.

51   Kitchens RL, Munford RS: CD14-dependent internalization of lipopolysaccharide (LPS) is strongly influenced by LPS aggregation but not by cellular responses to LPS. J Immunol 1998;160:1920–1928.

52   Gegner JA, Ulevitch RJ, Tobias PS: Lipopolysaccharide (LPS) signal transduction and clearance: Dual roles for LPS binding protein and membrane CD14. J Biol Chem 1995;270:5320–5326.

53   Hailman E, Lichenstein HS, Wurfel MM, Miller DS, Johnson DA, Kelley M, Busse LA, Zukowski MM, Wright SD: Lipopolysaccharide (LPS)-binding protein accelerates the binding of LPS to CD14. J Exp Med 1994;179:269–277.

54   Tobias PS, Soldau K, Gegner JA, Mintz D, Ulevitch RJ: Lipopolysaccharide binding protein-mediated complexation of lipopolysaccharide with soluble CD14. J Biol Chem 1995;270:10482–10488.

55   Hailman E, Vasselon T, Kelley M, Busse LA, Hu MCT, Lichenstein HS, Detmers PA, Wright SD: Stimulation of macrophages and neutrophils by complexes of lipopolysaccharide and soluble CD14. J Immunol 1996;156:4384–4390.

56   Yu B, Wright SD: Catalytic properties of lipopolysaccharide (LPS) binding protein: Transfer of LPS to soluble CD14. J Biol Chem 1996;271:4100–4105.

57   Perera PY, Qureshi N, Christ WJ, Stütz P, Vogel SN: Lipopolysaccharide and its analog antagonists display differential serum factor dependencies for induction of cytokine genes in murine macrophages. Infect Immun 1998;66:2562–2569.

58   Tapping RI, Tobias PS: Cellular binding of soluble CD14 requires lipopolysaccharide (LPS) and LPS-binding protein. J Biol Chem 1997;272:23157–23164.

59   Vasselon T, Pironkova R, Detmers PA: Sensitive responses of leukocytes to lipopolysaccharide require a protein distinct from CD14 at the cell surface. J Immunol 1997;159:4498–4505.

60   Juan TSC, Kelley MJ, Johnson DA, Busse LA, Hailman E, Wright SD, Lichenstein HS: Soluble CD14 truncated at amino acid 152 binds lipopolysaccharide (LPS) and enables cellular response to LPS. J Biol Chem 1995;270:1382–1387.

61   Viriyakosol S, Kirkland TN: The N-terminal half of membrane CD14 is a functional cellular lipopolysaccharide receptor. Infect Immun 1996;64:653–656.

62   Viriyakosol S, Kirkland TN: A region of human CD14 required for lipopolysaccharide binding. J Biol Chem 1995;270:361–368.

63 Stelter F, Bernheiden M, Menzel R, Jack RS, Witt S, Fan X, Pfister M, Schütt C: Mutation of amino acids 39-44 of human CD14 abrogates binding of lipopolysaccharide and *Escherichia coli*. Eur J Biochem 1997;243:100–109.

64 McGinley MD, Narhi LO, Kelley MJ, Davy E, Robinson J, Rohde MF, Wright SD, Lichenstein HS: CD14: Physical properties and identification of an exposed site that is protected by lipopoly-saccharide. J Biol Chem 1995;270:5213–5218.

65 Juan TSC, Hailman E, Kelley MJ, Busse LA, Davy E, Empig CJ, Narhi LO, Wright SD, Lichenstein HS: Identification of a lipopolysaccharide binding domain in CD14 between amino acids 57 and 64. J Biol Chem 1995;270:5219–5224.

66 Shapiro RA, Cunningham MD, Ratcliffe K, Seachord C, Blake J, Bajorath J, Aruffo A, Darveau RP: Identification of CD14 residues involved in specific lipopolysaccharide recognition. Infect Immun 1997;65:293–297.

67 Lee J-D, Kravchenko V, Kirkland TN, Han J, Mackman N, Moriarty A, Leturcq D, Tobias PS, Ulevitch RJ: Glycosyl-phosphatidylinositol-anchored or integral membrane forms of CD14 mediate identical cellular responses to endotoxin. Proc Natl Acad Sci USA 1993;90:9930–9934.

68 Pugin J, Kravchenko VV, Lee J-D, Kline L, Ulevitch RJ, Tobias PS: Cell activation mediated by glycosylphosphatidylinositol-anchored or transmembrane forms of CD14. Infect Immun 1998;66:1174–1180.

69 Kitchens RL, Wang P-Y, Munford RS: Bacterial lipopolysaccharide can enter human monocyte-macrophages via two CD14-dependent pathways. J Immunol 1998;161:5534–5545.

70 Gallay P, Jongeneel CV, Barras C, Burnier M, Baumgartner J-D, Glauser MP, Heumann D: Short time exposure to lipopolysaccharide is sufficient to activate human monocytes. J Immunol 1993;150:5086–5093.

71 Luchi M, Munford RS: Binding, internalization, and deacylation of bacterial lipopolysaccharides by human neutrophils. J Immunol 1993;151:959–969.

72 Detmers PA, Thiéblemont N, Vasselon T, Pironkova R, Miller DS, Wright SD: Potential role of membrane internalization and vesicle fusion in adhesion of neutrophils in response to lipopolysaccha-ride and TNF. J Immunol 1996;157:5589–5596.

73 Kitchens RL, Munford RS: Internalization of LPS by phagocytes; in Morrison DC, Brade H, Opal S, Vogel S (eds): Endotoxin in Health and Disease. New York, Marcel Dekker, 1999, pp 521–536.

74 Lichtman SN, Wang J, Zhang C, Lemasters JJ: Endocytosis and Ca^{2+} are required for endotoxin-stimulated TNF-a release by rat Kupffer cells. Am J Physiol [G] 1996;271:G920–G928.

75 Thieblemont N, Wright SD: Mice genetically hyporesponsive to lipopolysaccharide (LPS) exhibit a defect in endocytic uptake of LPS and ceramide. J Exp Med 1997;185:2095–2100.

76 Thieblemont N, Thieringer R, Wright SD: Innate immune recognition of bacterial lipopolysaccha-ride: Dependence on interactions with membrane lipids and endocytic movement. Immunity 1998;8:771–777.

77 Schromm AB, Brandenburg K, Loppnow H, Zähringer U, Rietschel ET, Carroll SF, Koch MHJ, Kusumoto S, Seydel U: The charge of endotoxin molecules influences their conformation and IL-6-inducing capacity. J Immunol 1998;161:5464–5471.

78 Morrison DC: Nonspecific interactions of bacterial lipopolysaccharides with membranes and mem-brane components; in Berry LJ (ed): Handbook of Endotoxin, vol 3: Cellular Biology of Endotoxin. Amsterdam, Elsevier, 1985, pp 1–25.

79 Wurfel MM, Wright SD: Lipopolysaccharide-binding protein and soluble CD14 transfer lipopoly-saccharide to phospholipid bilayers: Preferential interaction with particular classes of lipid. J Immu-nol 1997;158:3925–3934.

80 Wang P-Y, Kitchens RL, Munford RS: Bacterial lipopolysaccharide binds to low-density domains of the monocyte-macrophage plasma membrane. J Inflamm 1996;47:126–137.

81 Poussin C, Foti M, Carpentier JL, Pugin J: CD14-dependent endotoxin internalization via a macropi-nocytic pathway. J Biol Chem 1998;273:20285–20291.

82 Munford RS, Hall CL: Detoxification of bacterial lipopolysaccharides (endotoxins) by a human neutrophil enzyme. Science 1986;234:203–205.

83 Pohlman TH, Munford RS, Harlan JM: Deacylated lipopolysaccharide inhibits neutrophil adherence to endothelium induced by lipopolysaccharide in vitro. J Exp Med 1987;165:1393–1402.

84 Lynn WA, Golenbock DT: Lipopolysaccharide antagonists. Immunol Today 1992;13:271–276.
85 Ulmer AJ, Feist W, Heine H, Kirikae T, Kirikae F, Kusumoto S, Kusama T, Brade H, Schade U, Rietschel ET, Flad H-D: Modulation of endotoxin-induced monokine release in human monocytes by lipid A partial structures that inhibit binding of ^{125}I-lipopolysaccharide. Infect Immun 1992;60: 5145–5152.
86 Kawata T, Bristol JR, Rose JR, Rossignol DP, Christ WJ, Asano O, Bubuc G, Gavin WE, Hawkins LD, Kishi Y, McGuinness PD, Mullarkey MA, Perez M, Robidoux ALC, Wang Y, Kobayashi S, Kimura A, Katayama K, Yamatsu I: Anti-endotoxin activity of a novel synthetic lipid A analog. Prog Clin Biol Res 1995;392:499–509.
87 Heine H, Brade H, Kusumoto S, Kusama T, Rietschel ETh, Flad H-D, Ulmer AJ: Inhibition of LPS binding on human monocytes by phosphonooxyethyl analogs of lipid A. J Endotoxin Res 1994;1:14–20.
88 Rose JR, Christ WJ, Bristol JR, Kawata T, Rossignol DP: Agonistic and antagonistic activities of bacterially derived *Rhodobacter sphaeroides* lipid A: Comparison with activities of synthetic material of the proposed structure and analogs. Infect Immun 1995;63:833–839.
89 Christ WJ, Asano O, Robidoux ALC, Perez M, Wang Y, Dubuc GR, Gavin WE, Hawkins LD, McGuinness PD, Mullarkey MA, Lewis MD, Kishi Y, Kawata T, Bristol JR, Rose JR, Rossignol DP, Kobayashi S, Hishinuma I, Kimura A, Asakawa N, Katayama K, Yamatsu I: E5531, a pure endotoxin antagonist of high potency. Science 1995;268:80–83.
90 Strittmatter W, Weckesser J, Salimath PV, Galanos C: Nontoxic lipopolysaccharide from *Rhodopseudomonas sphaeroides* ATCC 17023. J Bacteriol 1983;155:153–158.
91 Loppnow H, Libby P, Freudenberg M, Krauss JH, Weckesser J, Mayer H: Cytokine induction by lipopolysaccharide (LPS) corresponds to lethal toxicity and is inhibited by nontoxic *Rhodobacter capsulatus* LPS. Infect Immun 1990;58:3743–3750.
92 Ingalls RR, Monks BG, Savedra R, Christ WJ, Delude RL, Medvedev AE, Espevik T, Golenbock DT: CD11/CD18 and CD14 share a common lipid A signaling pathway. J Immunol 1998;161: 5413–5420.
93 Munford RS, Hall CL: Uptake and deacylation of bacterial lipopolysaccharides by macrophages from normal and endotoxin-hyporesponsive mice. Infect Immun 1985;48:464–473.
94 Takakuwa T, Knopf HP, Sing A, Carsetti R, Galanos C, Freudenberg MA: Induction of CD14 expression in *Lps*[o], *Lps*[d] and tumor necrosis factor receptor-deficient mice. Eur J Immunol 1996; 26:2686–2692.
95 Espevik T, Otterlei M, Skjak-Braek G, Ryan L, Wright SD, Sundan A: The involvement of CD14 in stimulation of cytokine production by uronic acid polymers. Eur J Immunol 1993;23:255–261.
96 Medvedev AE, Flo T, Ingalls RR, Golenbock DT, Teti G, Vogel SN, Espevik T: Involvement of CD14 and complement receptors CR3 and CR4 in nuclear factor-kappaB activation and TNF production induced by lipopolysaccharide and group B streptococcal cell walls. J Immunol 1998; 160:4535–4542.
97 Savedra R Jr, Delude RL, Ingalls RR, Fenton MJ, Golenbock DT: Mycobacterial lipoarabinomannan recognition requires a receptor that shares components of the endotoxin signaling system. J Immunol 1996;157:2549–2554.
98 Weidemann B, Brade H, Rietschel ET, Dziarski R, Bazil V, Kusumoto S, Flad H-D, Ulmer AJ: Soluble peptidoglycan-induced monokine production can be blocked by anti-CD14 monoclonal antibodies and by lipid A partial structures. Infect Immun 1994;4709-4715.
99 Gupta D, Kirkland TN, Viriyakosol S, Dziarski R: CD14 is a cell-activating receptor for bacterial peptidoglycan. J Biol Chem 1996;271:23310–23316.
100 Weidemann B, Schletter J, Dziarski R, Kusumoto S, Stelter F, Rietschel ET, Flad HD, Ulmer AJ: Specific binding of soluble peptidoglycan and muramyldipeptide to CD14 on human monocytes. Infect Immun 1997;65:858–864.
101 Dziarski R, Tapping RI, Tobias PS: Binding of bacterial peptidoglycan to CD14. J Biol Chem 1998;273:8680–8690.
102 Cleveland MG, Gorham JD, Murphy TL, Tuomanen E, Murphy KM: Lipoteichoic acid preparations of gram-positive bacteria induce interleukin-12 through a CD14-dependent pathway. Infect Immun 1996;64:1906–1912.

103 Hattor Y, Kasai K, Akimoto K, Thiemermann C: Induction of NO synthesis by lipoteichoic acid from *Staphylococcus aureus* in J774 macrophages: Involvement of a CD14-dependent pathway. Biochem Biophys Res Commun 1997;233:375–379.

104 Kusunoki T, Hailman E, Juan TSC, Lichenstein HS, Wright SD: Molecules from *Staphylococcus aureus* that bind CD14 and stimulate innate immune responses. J Exp Med 1995;182:1673–1682.

105 Kusunoki T, Wright SD: Chemical characteristics of *Staphylococcus aureus* molecules that have CD14-dependent cell-stimulating activity. J Immunol 1996;157:5112–5117.

106 Sellati TJ, Bouis DA, Kitchens RL, Darveau RP, Pugin J, Ulevitch RJ, Gangloff SC, Goyert SM, Norgard MV, Radolf JD: *Treponema pallidum* and *Borrelia burgdorferi* lipoproteins and synthetic lipopeptides activate monocytic cells via a CD14-dependent pathway distinct from that used by lipopolysaccharide. J Immunol 1998;160:5455–5464.

107 Wooten RM, Morrison TB, Weis JH, Wright SD, Thieringer R, Weis JJ: The role of CD14 in signaling mediated by outer membrane lipoproteins of *Borrelia burgdorferi*. J Immunol 1998;160:5485–5492.

108 Yu B, Hailman E, Wright SD: Lipopolysaccharide binding protein and soluble CD14 catalyze exchange of phospholipids. J Clin Invest 1997;99:315–324.

109 Wang P-Y, Kitchens RL, Munford RS: Phosphatidylinositides bind to plasma membrane CD14 and can prevent monocyte activation by bacterial lipopolysaccharide. J Biol Chem 1998;273:24309–24313.

110 Devitt A, Moffatt OD, Raykundalia C, Capra JD, Simmons DL, Gregory CD: Human CD14 mediates recognition and phagocytosis of apoptotic cells. Nauture 1998;392:505–509.

111 Soell M, Lett E, Holveck F, Schöller M, Wachsmann D, Klein J-P: Activation of human monocytes by streptococcal rhamnose glucose polymers is mediated by CD14 antigen, and mannan binding protein inhibits TNF-α release. J Immunol 1995;154:851–860.

112 Zhang Y, Doerfler M, Lee TC, Guillemin B, Rom WN: Mechanisms of stimulation of interleukin 1b and tumor necrosis factor-α by *Mycobacterium tuberculosis* components. J Clin Invest 1993;91:2076–2083.

113 Peterson PK, Gekker G, Hu S, Sheng WS, Anderson WR, Ulevitch RJ, Tobias PS, Gustafson KV, Molitor TW, Chao CC: CD14 receptor-mediated uptake of nonopsonized *Mycobacterium tuberculosis* by human microglia. Infect Immun 1995;63:1598–1602.

114 Newman SL, Chaturvedi S, Klein BS: The W-1 antigen of *Blastomyces dermatitidis* yeasts mediates binding to human macrophage CD11b/CD18 (CR3) and CD14. J Immunol 1995;154:753–761.

115 Lipovsky MM, Gekker G, Anderson WR, Molitor TW, Peterson PK, Hoepelman AIM: Phagocytosis of nonopsonized *Cryptococcus neoformans* by swine microglia involves CD14 receptors. Clin Immunol Immunopathol 1997;84:298–211.

116 Beekhuizen H, Blokland I, Corsel-van Tilburg A, Koning F, Van Furth R: CD14 contributes to the adherence of human monocytes to cytokine-stimulated endothelial cells. J Immunol 1991;147:3761–3767.

117 Bosco MC, Espinoza-Delgado I, Rowe TK, Malabarba MG, Longo DL, Varesio L: Functional role for the myeloid differentiation antigen CD14 in the activation of human monocytes by IL-2. J Immunol 1997;159:2922–2931.

Richard L. Kitchens, PhD, Department of Internal Medicine,
University of Texas Southwestern Medical Center,
5323 Harry Hines Blvd., Dallas, TX 75235–9113 (USA)
Fax +1 (214) 648 9478, E-Mail rkitch@mednet.swmed.edu

Jack RS (ed): CD14 in the Inflammatory Response.
Chem Immunol. Basel, Karger, 2000, vol 74, pp 83–107

··········

Interactions of CD14 with Components of Gram-Positive Bacteria

Roman Dziarski [a], *Artur J. Ulmer* [b], *Dipika Gupta* [a]

[a] Northwest Center for Medical Education, Indiana University School of Medicine, Gary, Ind., USA;
[b] Division of Cellular Immunology, Research Center Borstel, Borstel, Germany

Discovery of CD14-Dependent Cell Activation by Non-Lipopolysaccharide Stimulants

Soon after the discovery of CD14 function as the main macrophage-activating receptor for gram-negative lipopolysaccharide (LPS) [1, 2], it was discovered that anti-CD14 mAbs could inhibit macrophage activation by other molecules, such as uronic acid polymers [3] and mycobacterial lipoarabinomannan [4]. Then, two independent studies demonstrated CD14 involvement in macrophage activation by the cell walls from several gram-positive bacteria [5] and purified staphylococcal peptidoglycan (PGN) [6]. Another group also demonstrated that both lipophosphoglycan from *Leishmania donovani* and soluble PGN bound to an LPS receptor that had characteristics of CD14 [7], thus further suggesting that CD14 could function as a receptor not only for LPS, but also for other microbial and nonmicrobial macrophage activators.

These findings gave impetus to more thorough studies of the function of CD14 as a 'pattern recognition receptor' that recognizes not only LPS, but also other bacterial (table 1) and nonbacterial (table 2) products.

CD14 as the Receptor for Gram-Positive Cell Walls

The first indication that CD14 may function as the receptor for gram-positive bacteria was inhibition by anti-CD14 Abs of gram-positive cell wall-induced activation of murine and human macrophage cell lines [5]. This effect

Table 1. CD14-dependent macrophage activators from bacterial cell walls

Cell wall compound	Bacterial source	Ref. No.
LPS	gram-negative bacteria	1
β1,4-*D*-Mannuronic acid (poly M)	*Pseudomonas syringae*	3
Peptidoglycan, insoluble	*Staphylococcus aureus*	8
Peptidoglycan, soluble	*Staphylococcus aureus*	6, 8–10
Lipoteichoic acid	gram-positive bacteria	11–13
Cell walls, insoluble	gram-positive bacteria	5, 8, 14, 15
Rhamnose-glucose polymer	Streptococcus	16
Lipoarabinomannan	Mycobacteria	4, 5, 8, 17
Lipoprotein, lipopeptide	Spirochetes	18, 19

Table 2. CD14-dependent nonbacterial activators

Compound	Source	Ref. No.
Poly uronic acid	*Ascophyllum nodosum*	3
Chitosans	arthropods	20
WI-1 (cell wall Ag)	*Blastomyces dermatitidis*	21
Fucoidan	*Fucus vesiculosus*	22
β1,4-Glucuronic acid	synthetic	3
Phospholipids	mammalian, synthetic	23, 24
Taxol	*Taxus brevifolia*	25
IL-2	T lymphocytes	26
Apoptotic cell membranes	mammalian cells	27

was seen with the cell walls from several gram-positive bacteria, including *Staphylococcus aureus, Bacillus subtilis, Streptococcus* group A and B, and *Streptococcus pneumoniae.*

   More definitive evidence that CD14 functions as the cell-activating receptor for gram-positive cell walls came from experiments with CD14 transfectants. 70Z/3 cells, which are CD14-negative immature mouse B cells, only respond to high concentrations of gram-positive cell walls. However, transfection of these cells with CD14 and expression of either GPI-linked or transmembrane forms of CD14 makes these cells more than 100 times more sensitive to activation by gram-positive cell walls [5].

CD14-dependence of macrophage activation by streptococcal [8, 16] and pneumococcal [15] cell walls, or streptococcal rhamnose-glucose polymer [14] was also later confirmed by other investigators.

CD14 as the Receptor for Peptidoglycan

Structure and Biologic Activity of Peptidoglycan
Peptidoglycan (PGN) is found in the cell walls of all bacteria and is especially abundant in the cell walls of gram-positive bacteria. PGN is composed of a glycan backbone of up to 100 alternating units of β-$(1 \rightarrow 4)$-linked N-acetylglucosamine (GlcNAc) and N-acetylmuramic acid (MurNAc), with short peptides linked to the lactyl group of the MurNAc residues (fig. 1A) [28]. Crosslinking of these peptides yields an enormous basket-like macromolecule surrounding the cytoplasmic membrane. Soluble polymeric uncross-linked PGN can also be secreted from gram-positive bacteria when they grow in the presence of β-lactam antibiotics, due to inhibition of peptide crosslinking and lack of incorporation of newly synthesized PGN into the existing cell wall [29]. In the cell wall, numerous macromolecules, such as cell wall teichoic acid, polysaccharides, and proteins are often covalently bound to PGN [29].

Despite quite different chemical structure from LPS (fig. 1C), PGN has many similar biologic activities [30–32], although usually it is not as active as LPS. This large number and similarity of biologic effects of LPS and PGN is due to the release of similar mediators (TNF-α, IL-1, IL-6, etc.) from host cells, which accounts for the ability of LPS and PGN to reproduce all major clinical manifestations of bacterial infections, including fever, inflammation, hypotension, leukocytosis, sleepiness, decreased appetite, malaise, and arthritis [32].

Function of CD14 as the Peptidoglycan Receptor
Simultaneously with and independently of the discovery of the function of CD14 as the receptor for gram-positive cell walls, it was also discovered that CD14 serves as the receptor for staphylococcal PGN. The first indication that CD14 serves as the PGN receptor came from the experiments showing that anti-CD14 mAbs inhibit PGN-induced production of cytokines (IL-1 and IL-6) in human monocytes [6]. Moreover, synthetic LPS partial structures that act as LPS antagonists (compounds 406 and 606), also inhibited PGN-induced activation of cytokine secretion by PGN in human monocytes [6].

More evidence for the function of CD14 as a PGN receptor was provided by the experiments with CD14 transfectants. 70Z/3 cells, which are CD14-negative immature mouse B cells, do not respond to PGN. However, 70Z/3-hCD14 transfectants (70Z/3 cells stably transfected with human CD14) re-

Soluble peptidoglycan
from *S. aureus*
- [→4)-ß-D-GlcNAcp-(1→4)-ß-D-MurNAcp- 3→
→L-Ala→D-Gln→L-Lys-[(Gly)₅]→D-Ala→D-Ala-(1-]ₙ⁻

1A

Lipoteichoic acid from *S. aureus*
R = D-Ala, GlcNAc, or H
Gro(2-R)3-P-[→1Gro(2-R)3-P-]₂₄₋₃₄→
→6-D-Glcpß1→6-D-Glcpß1→1Gro(2,3-diAcyl₁₄₋₁₉)

1B

spond to PGN, as evidenced by activation of a ubiquitous transcription factor NF-κB, accompanied by degradation of its inhibitor IκB-α, followed by differentiation into surface IgM-expressing B cells [8]. Moreover, activation of 70Z/3-hCD14 transfectants by PGN (similarly to LPS) is inhibited by anti-CD14 mAbs [8].

Binding of Peptidoglycan to CD14

The first indication that PGN binds to CD14 came from experiments showing that specific binding of PGN to human monocytes was inhibited by

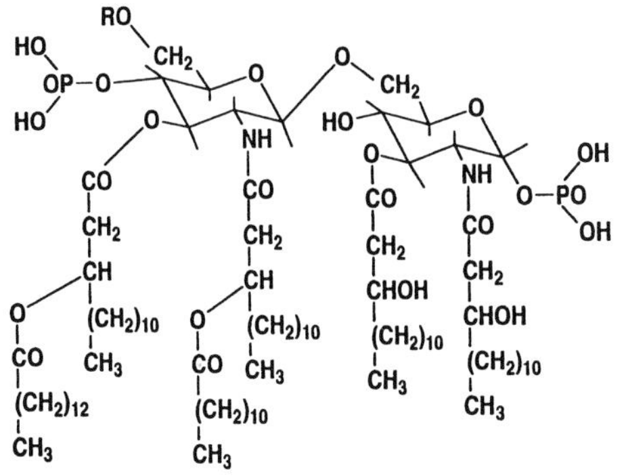

Lipid A, R = H
Re-LPS, R = [(2→4)-α-KDO]₃-(2-
LPS, R = inner core + outer core + O-Ag
from *Enterobacteriaceae*

1C

Fig. 1. Chemical structures of uncrosslinked *S. aureus* PGN (*A*), *S. aureus* LTA (*B*), and *Enterobacteriaceae* lipid A, ReLPS, and LPS (*C*).

anti-CD14 mAbs [9]. PGN binding was also inhibited by LPS and an LPS antagonist, compound 406 [9]. Specific binding of PGN to membrane CD14 (mCD14) was then confirmed by photoaffinity crosslinking and immunopre-cipitation of PGN-CD14 complexes with anti-CD14 mAbs [10].

PGN also binds to soluble CD14 (sCD14) at a ratio of PGN to sCD14 of approximately 1:1, as demonstrated by a difference in electrophoretic mobility between sCD14 and PGN:sCD14 complexes in native electrophoresis [9]. Binding of sCD14 to PGN was also confirmed by three other assays: binding of ^{32}P-labeled sCD14 to agarose-immobilized PGN, photoaffinity cross-link-ing, and binding of biotin-labeled PGN to immobilized sCD14 in an ELISA assay [10].

Binding of sCD14 to PGN (immobilized to agarose) is slower than to immobilized ReLPS but of higher affinity ($K_d = 25$ nM for PGN versus 41 nM for ReLPS) [10]. LPS-binding protein (LBP) increases the binding of sCD14 to PGN by adding another lower affinity K_d and another higher B_{max}, in contrast to ReLPS, for which LBP increases the affinity of binding by yielding two K_d with significantly higher affinity (7 and 27 nM). LBP also enhances inhibition of sCD14 binding to immobilized PGN and LPS by soluble LPS, ReLPS, and lipid A, but not by soluble PGN. Therefore, LBP enhances the binding of PGN and LPS to CD14 by different mechanisms. For LPS, it increases the affinity of binding and lowers the cell-activating concentration of LPS. By contrast, for PGN, LBP only increases its low-affinity binding [10], which does not enhance CD14-dependent cell activation by PGN [6, 32].

Polymeric Peptidoglycan Is Required for Binding to CD14 and Cell Activation

The exact structural requirements of PGN for binding to CD14 have not yet been determined. It is known, however, that polymeric PGN is needed both for the binding [10] and cell activation [33]. CD14 binds to insoluble polymeric PGN and to soluble polymeric PGN, but does not bind to soluble synthetic or natural PGN fragments, such as monomeric muramyl dipeptide (MDP), or PGN pentapeptide, or dimeric GlcNAc-MDP [10]. Moreover, digestion of PGN with PGN-lytic enzymes reduces the binding of PGN to CD14 [10] and PGN-induced cell activation [33] proportionately to the extent of digestion. CD14, however, binds to agarose-immobilized MDP or GlcNAc-MDP [10], but not to agarose-immobilized PGN pentapeptide. These results suggest that solid-phase-bound MDP or GlcNAc-MDP mimic the CD14-binding polymeric PGN structure, and also indicate that the glycan part of PGN (but not the entire peptide) is essential for the binding to CD14.

The Regions of CD14 Involved in Binding to Peptidoglycan and Cell Activation

Similarly to LPS, less than half of the CD14 molecule, i.e. the N-terminal 152 amino acids, are sufficient both for PGN binding (with similar affinity as the full length CD14) and for CD14-mediated cell activation by PGN [8, 10]. These results indicate that both the binding and cell-activation domains for both LPS and PGN are located within the N-terminal 152 amino acid fragment of CD14.

The exact binding sites for both LPS [35, 36] and PGN [10] seem to be conformational. The sequences that are most critical for CD14 binding to both LPS and PGN are located between amino acids 51–64 (the binding site of anti-CD14 mAb MEM-18), because anti-CD14 mAb MEM-18 is most efficient in inhibiting CD14 binding to both ReLPS (by over 95%) and to PGN (by over 80%). MEM-18 is also most efficient (out of 14 anti-CD14 mAbs) in inhibiting cell activation by both LPS and PGN, indicating that the same epitope on CD14 is of primary importance for both binding and cell activation by both LPS and PGN.

This region, however, may not be sufficient for binding of LPS and PGN, because mAbs specific to other regions of CD14 also partially inhibit binding of LPS and PGN and cell activation [10, 36]. However, these other sequences that contribute to LPS and PGN binding and cell activation are at least partially different, because there are several anti-CD14 mAbs (directed to more N-terminal regions of CD14) that inhibit LPS binding but not PGN binding, and one mAb (directed to a more C-terminal region of CD14) that inhibits PGN binding but not LPS binding [10]. Therefore, it appears that LPS and PGN bind to conformational rather than linear CD14 epitopes that are partially similar (amino acids 51–64) and partially different.

Similarly, the domains of CD14 most critical for mCD14-mediated cell activation by both LPS and PGN are located within the N-terminal 65 amino acids [8]. However, the specific amino acid sequences responsible for cell activation by LPS and PGN are not identical, since some mCD14 deletion mutants are still responsive to PGN, but unresponsive to LPS [8].

CD14 as the Receptor for Lipoteichoic Acids

Lipoteichoic acids (LTA) are amphiphilic glycolipids (fig. 1B) present in the cell walls of gram-positive, but not gram-negative, bacteria. LTA can activate macrophages to secrete cytokines or nitric oxide, and in vivo, they also act synergistically with PGN to induce multiple organ failure and shock [37, 38]. LTA induce CD14-dependent secretion of IL-12 in human cells, which is inhibited by anti-CD14 mAb MY-4, as well as by an LPS antagonist, *Rhodobacter sphaeroides* LPS [11]. LTA also induce NO synthase in mouse macrophages in a CD14-dependent manner [13].

However, the exact chemical structure responsible for biological activity of LTA is still not known [39], because chemically synthesized structures resembling LTA of *Enterococcus hirae* were not active in inducing cytokines and antitumor activity. Therefore, it appears that the biological activity of purified natural LTA is not due to the main constituent of the LTA preparations [40].

This conclusion is supported by the finding that the CD14-dependent cytokine-inducing factor in *S. aureus* LTA preparation, fractionated on a reverse-phase column, was distinct from the main LTA fraction, and the main purified LTA fraction failed to stimulate IL-6 release in human monocytes and U373 cells [12]. However, this main purified LTA fraction could block the sCD14-dependent activation of U373 cells by LPS and also could bind directly to CD14, as shown by a shift in the electrophoretic mobility of sCD14 in native electrophoresis [12]. Moreover, LTA also inhibit binding of both LPS and PGN to CD14 [10]. Therefore, LTA preparations are heterogeneous and contain both agonistic and antagonistic LTA-like molecules.

CD14 as the Receptor for Mycobacterial Lipoarabinomannan

The lipoglycan lipoarabinomannan (LAM), the major antigen of myco-bacterial cell walls, stimulates cytokine production in human and murine monocytes and macrophages. In 1993, it was found that both anti-CD14 mAb and lipid IV_A inhibit LAM-induced cytokine release in human THP-1 cell line [4]. Later it was also shown that transfection of 70Z/3 cells with CD14 makes these cells responsive to LAM [5, 8]. LAM that lacks terminal mannosyl units is reactive with CD14, whereas, terminally mannosylated LAM is a poor activator of cytokines and preferentially binds to the macrophage mannose receptor [41]. Similarly to LPS and PGN, 152 N-terminal amino acids of CD14 are also sufficient for the responsiveness to LAM [42]. However, the mechanisms of cell activation by LAM and PGN are not identical, because, in contrast to PGN [6, 33], monocyte and premonocyte activation by LAM is enhanced by sCD14 and LBP [17, 42].

LAM directly interacts with CD14, as shown by inhibition of changes in fluorescence intensity of FITC-LPS induced by the binding to sCD14 [5].

CD14 as the Receptor for Other Bacterial and Nonbacterial Polymers

In addition to LPS, PGN, LTA, and LAM, other bacterial cell wall compounds, such as lipopeptides, sphingolipids, and also nonbacterial and synthetic polymers induce production of cytokines in monocytes and macro-phages, and several of these compounds can stimulate these cells in a CD14-dependent manner (tables 1, 2). It should be noted, however, that CD14 is not involved in the stimulation of macrophages by every cell wall component, because, for example, glycosphingolipid from *Sphingiomonas paucibilis* induces CD14-independent cytokine production in human monocytes [43].

Differential Activation of CD14-Negative Cells by Complexes of Soluble CD14 with Bacterial Cell Wall Components

Normal human serum contains 4–6 µg/ml of sCD14, which lacks the GPI anchor, but has the same amino acid sequence as mCD14. Cells that do not express mCD14, such as vascular endothelial cells, epithelial cells, vascular smooth muscle cells, fibroblasts, and astrocytes can be activated by complexes of LPS with sCD14 (table 3). Formation of sCD14:LPS complexes is greatly enhanced by LBP and, thus, activation of endothelial and other mCD14-negative cells by LPS and sCD14 is greatly enhanced by LBP [45–50].

Table 3. Activation of CD14-negative cells by bacterial cell wall components and sCD14 or transfected mCD14

Compound (source)	Cell line (cell type)	Stimulation through		Ref. No.
		sCD14	mCD14	
LPS (gram-negative bacteria)	70Z/3 (mouse pre-B cell)		yes	5, 44
	HL-60 (myelocytic)	yes		42
	endothelial cells	yes	yes	45–51
	vascular smooth muscle	yes		50
	ECV304 (endothelial)	yes		52
	U373 (epithelial)	yes	yes	46, 51
	SW620 (epithelial)	yes	yes	45
	CHO (hamster fibroblast)		yes	17, 53, 54
	HT1080 (fibroblast)		yes	17, 53
	293 (embryonic kidney)		no	55, 56
Cell walls (B. subtilis)	70Z/3 (mouse pre-B cell)		yes	5
	SW620 (epithelial)	yes		5
Cell walls (S. pneumoniae)	astrocytes	yes		57
Cell walls (group B. Streptococcus)	CHO (hamster fibroblast)		yes	16
PGN (S. aureus)	70Z/3 (mouse pre-B cell)		yes	8
	HUVEC (endothelial)	no		52
	ECV304 (endothelial)	no		52
	U373 (epithelial)	no		52
	CHO (hamster fibroblast)		no	52
	HT1080 (fibroblast)		no	52
	293 (embryonic kidney)		no	67a
LAM (Mycobacteria)	70Z/3 (mouse pre-B cell)		yes	5, 8
	HL-60 (myelocytic)	yes		41
	endothelial cells	no	yes	51
	U373 (epithelial)	no	yes	17, 51
	SW620 (epithelial)	no	yes	45, 51
	HT1080 (fibroblast)		no	17
	CHO (hamster fibroblast)		no	17
Lipoprotein (B. burgdorferi, T. pallidum)	endothelial cells	yes		19
	CHO (hamster fibroblast)		no	18
Lipopeptide (synthetic)	CHO (hamster fibroblast)		no	18
Poly-uronic acid (A. nodosum, P. syringae, synthetic)	U373 (epithelial)	no		3

The exact mechanism of activation of these mCD14-negative cells by sCD14:LPS complexes and the binding sites for these complexes on CD14-negative cells are still not known. Although identification of binding sites for sCD14-LPS complexes on endothelial cells has been reported [58], these results could not be confirmed by other investigators, who could detect LBP-dependent binding of sCD14:LPS:LBP complexes to nonmyeloid mCD14-negative cells, but this binding led to LPS internalization and was *not* directly involved in cellular activation [59].

Vascular endothelial cells participate in inflammation, organ failure, and shock by being both a producer of and a target for proinflammatory mediators. Stimulation of endothelial cells by sCD14:LPS complexes induces production of proinflammatory cytokines (e.g. IL-1 and IL-6) and chemokines (e.g. IL-8) and expression of adhesion molecules. These adhesion molecules promote inflammation by enhancing attachment of leukocytes to vascular endothelium and extravasation of leukocytes and their migration into the inflamed tissues [60, 61].

Because vascular endothelial and other mCD14-negative cells are also strongly activated by proinflammatory cytokines, such as TNF-α and IL-1, these cells are activated much more efficiently by LPS in the presence of small amounts of whole blood (2–4%) than in the presence of serum or purified sCD14 and LBP. This indirect activation is mediated by TNF-α and IL-1 that originate from LPS-activated monocytes and macrophages [62, 63]. It is likely, therefore, that in vivo, similarly to ex vivo, indirect activation of endothelial cells by cytokines induced by LPS from monocytes and macrophages is of primary importance.

Whole gram-positive cell walls and spirochetal lipoproteins, similarly to LPS, can activate CD14-negative endothelial and epithelial cells in the presence of sCD14 [5, 18, 19, 57] (table 3). PGN and LAM also bind to sCD14 with high affinity [5, 10] and form stable complexes with sCD14 [5, 9, 10]. However, sCD14:PGN and sCD14:LAM complexes do not activate endothelial and epithelial cells to secrete cytokines, to express adhesion molecules, or to activate NF-κB [17, 45, 46, 52] (table 3), even though LAM can activate these cells if they are made to express mCD14 through transfection [51] and sCD14:LAM complexes can activate CD14-negative premonocytic HL-60 cells [42]. On the other hand, fibroblast-like cells (CHO and HT1080) made to express mCD14 through transfection do not respond to LAM, PGN, lipoproteins, or lipopeptides, although they are highly responsive to LPS (table 3). However, there are also some cells (human embryonic kidney 293 cells) that are unresponsive even to LPS when transfected with CD14 [55, 56].

These findings underscore differences in the function of sCD14 and mCD14 as the facilitators of cell activation by LPS and various other bacterial CD14

---

ligands. The molecular and biochemical basis of these differences is still not understood, although differences in the expression and/or function of signal-transducing coreceptors or cell-activating molecules (such as the recently discovered Toll-like receptors) [55, 56] in different cells may be responsible for this phenomenon.

Even though several of the gram-positive cell wall components do not activate endothelial and epithelial cells directly, PGN, similarly to LPS, can induce very strong activation of these cells indirectly in the presence of even small amounts (2–4%) of whole blood [52]. Again, similarly to LPS, the secretion of both TNF-α and IL-1 from blood monocytes is responsible for this activation [52].

Function of CD14 as a 'Pattern-Recognition Receptor'

Because of the large number of structurally different ligands that can bind to CD14, three models for the function of CD14 as a cell-activating receptor have been proposed. According to the first model [5], CD14 serves as a 'pattern recognition receptor' that can recognize shared features of microbial cell surface components and can enable host cells to respond to pathogenic bacteria, but not to a great variety of other nonpathogenic or nonmicrobial polysaccharides. This model implied that CD14 can discriminate between different ligands and can control the specificity of macrophage responses. How this discrimination was achieved, however, was not clear. The second alternative model [64] proposed that CD14 does not have the recognition specificity and merely serves as an albumin-like carrier molecule that transfers ligands to an as yet unidentified recognition/cell activating molecule(s). The latter model was supported by the inability of CD14 to discriminate between agonistic and antagonistic derivatives of LPS [54]. Finally, in the third 'combinatorial' model, both CD14 and the putative recognition/cell-activating molecule would contribute to the specificity of cell activation. Other models, which proposed that CD14 could be a component of a heteromeric receptor complex and not directly bind other ligands, are less likely because of direct high affinity binding of PGN and other non-LPS ligands to CD14 [10].

It is still not possible to definitively discriminate among the first three models. However, it seems that CD14 has at least some specificity, because it binds some molecules with high affinity (e.g. LPS, PGN, LTA) and does not bind other similar molecules (e.g. ribitol teichoic acid, dextran, dextran sulfate, or heparin) [10]. These findings at least partially support CD14 function as a 'pattern recognition receptor'. It is still not clear, however, what are the chemical features of the 'pattern' that CD14 recognizes. It seemed that most ligands recognized by CD14 have polymeric carbohydrates often with closely located carbonyl residues (such as LPS or PGN). However, not all CD14 ligands have these structural features, because CD14 also binds ligands that do not have carbohydrates, such as lipoproteins [18, 19] or phospholipids [23, 24]. Therefore, it can be postulated that CD14 recognizes glycoconjugates or phospholipids via distinct patterns of ionic charges [65]. The specificity for these ligands is still tightly controlled, because several other charged polymers, e.g. dextran sulfate or heparin, do not bind to CD14 with affinity similar to LPS or PGN [10].

Recent results also suggest that recognition of different patterns is encoded in somewhat different regions of CD14, and that the binding sites are conformational and composed of several regions, partially identical and partially different for different ligands [10]. Such a multifunctional binding site could then much more easily accommodate specific binding to a variety of structurally different ligands.

Mechanism of Cell Activation by CD14

mCD14 is a glycosylphosphatidylinositol (GPI)-linked molecule that does not have a transmembrane domain. Therefore, by itself, it presumably cannot transmit the activating signal into the cell. For this reason, it has been proposed that either CD14:ligand complexes associate with other molecules, or CD14 ligands are transferred from CD14 to other cell-activating or coreceptor molecule(s) [45–47, 55, 56, 67].

The most likely candidates for this CD14 LPS coreceptor are the recently identified Toll-like receptors, TLR-2 and TLR-4, which are type I transmembrane proteins with leucine-rich repeats in their extracellular domains, and cytoplasmic domains with sequence homology to the IL-1 receptor. TLR-2 protein binds LPS, expression of TLR-2 converts LPS-unresponsive into LPS-responsive cells, the responsiveness is enhanced by coexpression of CD14, and some cells expressing CD14 alone and no TLR-2 are unresponsive to LPS [55, 56]. Moreover, LPS-low responder C3H/HeJ mice have an inactivating point mutation in the TLR-4 gene, which seems to be responsible for their low responsiveness to LPS [67]. It is still not clear how TLRs interact with CD14 and how they transmit LPS-induced signal into the cell.

It is also still not known if TLRs are needed for CD14-mediated responses to all CD14 ligands in all CD14-positive cells, although recent data demonstrate that TLR-2 (but not TLR-1 and TLR-4) also participates in the responsiveness to gram-positive bacteria and their cell wall components, PGN and LTA [67a].

GPI-linked cell-surface molecules accumulate together with cholesterol and glycosphingolipids forming special membrane microdomains called *rafts* or 'detergent insoluble glycolipid-enriched domains' (DIG) [68–70]. These domains are the sites of numerous cell functions from membrane traffic and cell morphogenesis to cell signaling, and they contain a high concentration of signaling molecules. Indeed, mCD14 was shown to be present in DIGs, associated mainly with the protein tyrosine kinase p53/56[lyn] [71, 72], but also with GTP-binding proteins and ouabain-inhibitable Na^+/K^+ ATPase in low-density domains of the monocyte membrane [73]. The GPI anchor of mCD14 was recently demonstrated to be responsible for the presence of the molecule in DIGs, since a transmembrane form of CD14 was not detected in the Triton X-100 insoluble cellular fraction [74]. However, the LPS-induced cellular responses mediated by transmembrane CD14 and GPI-linked CD14 are very similar [74]. So far it is not clear how mCD14 and other GPI-anchored molecules can be associated with tyrosine kinases that are present inside the cell or on the inner surface of the cell membrane. Furthermore, it is not known whether mCD14 associates with Lyn through TLRs.

Although there is no doubt that CD14 serves as a macrophage-activating receptor for components of gram-positive bacteria, the in vivo role of CD14 in gram-positive infections may be different from simply serving as a cell-activating receptor. Recent data indicate that CD14 does not play a role in enhancing the clinical manifestations of septic shock, because the lethality of CD14 knockout mice from *S. aureus* bacteremia is not greater than of control CD14-expressing mice [75]. Also surprisingly, the serum concentration of TNF-α following *S. aureus* injection was much higher in CD14-knockout than in CD14-expressing control mice, suggesting that CD14 may actually downregulate the cytokine responses to gram-positive bacteria in vivo [75]. The mechanism of this downregulation is not yet known, but it could be due to CD14-mediated enhancement of phagocytosis.

Indeed, in addition to triggering cell activation, CD14 also promotes internalization of its ligands. mCD14-bound ligands (including LPS and PGN), as well as sCD14:LPS complexes that bind to cells, are rapidly (within minutes) internalized [59, 76; Ulmer, unpubl.]. CD14 also mediates uptake of entire bacteria [77–80]. However, this internalization (at least for LPS) is independent of and not needed for cell activation, because there are some mAbs that inhibit cell activation but not LPS internalization, and some other mAbs that inhibit LPS internalization but not activation [59, 81]. Moreover, highly aggregated LPS is preferentially internalized as compared to less aggregated or monomeric LPS, whereas, aggregation does not enhance LPS-induced cell activation [82], and monomeric LPS is the preferred cell-activating species of LPS [83].

CD14-Independent Cell Activation by Gram-Positive Bacteria: Comparison with LPS

CD14 is undoubtedly the main macrophage and neutrophil receptor for cell activation and induction of cytokine production by low concentrations of LPS, because CD14-knockout mice are more than 1,000 times less sensitive to LPS than their wild-type littermates [84]. However, there are also CD14-independent mechanisms of cell activation by LPS, because cells from CD14-knockout mice do respond to high concentrations of LPS with production of cytokines [25]. Moreover, CD14-deficient mice show no alteration in the induction of acute phase proteins in vivo by LPS or lipid A [85]. Therefore, some effects of LPS are CD14-dependent, but other effects can be partially or totally CD14-independent.

The mechanism of these CD14-independent responses to LPS, however, is still poorly understood. As mentioned in the preceding section, TLR-2 and possibly TLR-4 can function as LPS signal-transducing molecules as well as LPS receptors without the presence of CD14 [55, 56, 67]. β2-Integrins, the complement receptors CR3 (CD11b/CD18) and CR4 (CD11c/CD18), can act as cell-activating LPS receptors, because transfection of CR3 or CR4 cDNA into CR3- and CR4-negative CHO cells confers upon them the ability to respond to LPS [86, 87]. However, neither CD14 [54, 88] nor CD11/CD18 [88] can discriminate between agonistic and antagonistic LPS analogs, which again underscores the importance of another signal-transducing molecule (TLR?) in LPS-induced cell activation. Transfection of CHO cells with CD55 (also known as decay-accelerating factor, DAF, a GPI-linked cell surface molecule that protects cells from complement-mediated damage) also makes these cells responsive to LPS, which indicates that CD55 (in addition to TLRs and β2-integrins) can serve as a CD14-independent LPS receptor [89]. However, the relative importance of β2-integrins or CD55 in LPS-induced responses, is still not known.

Even less is known about the mechanism of CD14-independent cell activation by other CD14 ligands. These ligands, e.g. PGN or whole gram-positive bacteria, can activate cells through CD14-independent pathways, because cells from CD14-knockout mice are only 5–10 times less sensitive to PGN than the cells from CD14-expressing wild-type littermates [Goyert and Dziarski, unpubl.]. The responses to whole pneumococcal bacteria are also CD14-independent [15], and in some systems responses to gram-positive cell wall components could not be inhibited with anti-CD14 Abs [90].

So far, however, no other receptors for PGN or other components of gram-positive bacteria have been found, although recent data indicate a role for TLR molecules [67a]. It is not known if CR3 and CR4 serve as receptors for

PGN, but by themselves they do not function as receptors for streptococcal cell walls, even though they may enhance CD14-mediated responses to these cell walls [16].

Signal Transduction Pathways Activated by Gram-Positive Bacterial Components: Comparison with LPS

Little is known about signal transduction pathways activated by components of gram-positive bacteria. Therefore, in this section, we will review CD14-dependent signal transduction pathways activated by LPS and PGN in their primary target cells, macrophages (fig. 2). Three points, however, should be noted. First, the CD14-dependent signal transduction pathways are likely to be actually initiated by an as yet unidentified coreceptor (possibly TLRs, see above), that may or may not be the same for LPS and PGN. Second, some effects may be initiated through CD14-independent mechanisms, and the receptors involved here may or may not be the same as the putative CD14 coreceptor. Third, in different cells different signal transduction pathways may be activated.

The earliest intracellular signal transduction event that can be detected within 1 or 5 min of macrophage activation by LPS or PGN, respectively, is increased tyrosine phosphorylation and activation of Lyn, and, in the case of LPS, possibly two other members of the Src family of tyrosine kinases, Hck and Fgr [72, 91, 92] (fig. 2A, B).

It is still not clear which signal transduction pathways are activated by Lyn in LPS- and PGN-stimulated cells. Src kinases often activate Syk or Vav, which can in turn lead to the activation of a small GTP-binding protein, Ras, which can then activate Raf, which then triggers activation of mitogen-activated protein kinases (MAP kinases). Such a connection of Lyn to the Ras→Raf→MAP kinase pathway has not so far been convincingly shown for LPS, even though LPS can induce increased tyrosine phosphorylation of Vav [93] and two other proteins (p145 and Shc) that associate with Syk [94]. However, activation of Syk by LPS has not been shown, and Syk-deficient cells show normal responses to LPS [95]. Activation of Lyn by LPS activates phosphatidylinositol 3-kinase (PI3-kinase) in a CD14-dependent manner [96], which in turn activates protein kinase C (PKC)-ζ [97]. PKC can activate the Raf→Mek→ERK pathway, but this link has not been so far shown for LPS. None of these pathways have been so far shown for PGN or other components of gram-positive bacteria.

Even though Lyn and other Src kinases may participate in the cell activation by LPS, and even though the Hck⁻/Fgr⁻ double knockout mice are resistant to endotoxic shock [98], these Src kinases are *not* required for several major effects of LPS, because macrophages from Lyn⁻/Hck⁻/Fgr⁻ triple knockout mice have no major defects in LPS-induced stimulation of nitrate production, IL-1, IL-6, and TNF-α secretion, as well as activation of ERK1/2 and JNK MAP kinases and transcription factor NF-κB [99]. The higher resistance to endotoxic shock of Hck⁻/Fgr⁻ double knockout mice could be due to the requirement of Hck and Fgr for the action of endotoxin-induced mediators, such as cytokines.

Both LPS and PGN activate MAP kinases. However, LPS strongly activates all three families of MAP kinases (ERK1/2, p38, and JNK) [91, 100–106] (fig. 2A), whereas, PGN strongly activates only ERK1/2 and JNK, and only marginally activates p38 [91, 105] (fig. 2B). Pneumococcal cell walls, however, activate both ERK and p38 [57].

MAP kinases are activated by dual phosphorylation on one threonine and one tyrosine by specialized kinases, called Mek or MKK (MAP kinase kinase), that show a high degree of specificity for individual MAP kinases. Thus, in several cells, including LPS-activated macrophages, Mek 1 and Mek 2 activate ERK 1 and ERK 2, Mek 3 and Mek 6 activate p38, and Mek 4 and Mek 7 activate JNK (although Mek 4 can also to some extent activate p38) [104, 106–112] (fig. 2A). It is not known which Meks activate which MAP kinases in PGN-stimulated cells (fig. 2B).

Although Meks can be activated by a variety of different Mek kinases (Mekk or MKKK), little is known about the mechanisms of LPS- and PGN-induced Mek activation. The Ras→Raf1→Mek1/2→ERK1/2 pathway has been proposed to be activated by LPS [107, 108, 113], but it is not known what activates Ras. However, there are also some indications that the Ras→Raf is not the main activating pathway for ERK1/2 [105, 109] or that the Ras→Raf pathway is not activated at all by LPS [114]. PKC, whose phorbol-insensitive forms (ε and ζ) are activated by LPS [97, 115], could also result in the activation of Mek1/2→ERK1/2 pathway, although this connection has not yet been established for LPS.

The significance of the activation of MAP kinases by LPS and PGN is still not clear, even though MAP kinases induce numerous transcription factors that can potentially induce numerous genes. For example, on one hand, inhibitors of ERK phosphorylation and activation, such as tyrphostins, block LPS-induced production of NO and TNF-α and prevent endotoxin lethality in vivo [116]. But on the other hand, strong activators of ERK, such as phorbol esters or CSF, do not mimic LPS or PGN effects, e.g. they induce very little TNF-α production and are not toxic [91, 117]. Similarly, selective activation of the Raf1→ERK pathway by a chimeric Raf1:estrogen receptor molecule induces, similarly to phorbol esters, only a small amount of TNF-α [113]. Also, activation of p38 MAP kinase alone cannot be responsi-

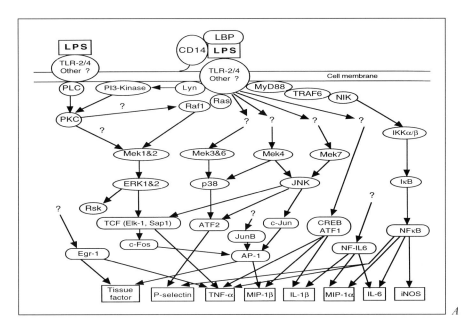

A

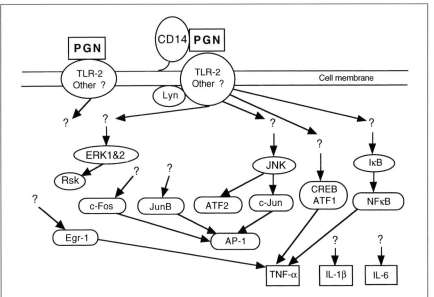

B

Fig. 2. Signal transduction pathways, transcription factors, and main genes activated by LPS (*A*) and PGN (*B*).

ble for or required for LPS- or PGN-induced cytokine production, because PGN, in contrast to LPS, does not effectively activate p38 [105], yet both PGN and LPS induce large amounts of TNF-α or IL-6 in macrophages [6, 91]. Therefore, it appears that activation of individual MAP kinases is not sufficient, or in some cases may not be even necessary, for the full induction of LPS- or PGN-stimulated genes.

In summary, it is likely that LPS and gram-positive products, including PGN, activate multiple signal transduction pathways that are mostly overlapping but partially different. The full spectrum of cell activation is likely to be the result of concerted action of all these pathways.

Transcription Factors and Genes Activated by Gram-Positive Bacterial Components: Comparison with LPS

Activation of a large number of transcription factors is induced by LPS, including NF-κB, NF-IL6, the CREB/ATF1 family, the AP-1 family, the Ets family (which includes Ets, Elk, Erg, and PU.1), and Egr (fig. 2A) [117–119]. Activation of several of these transcription factors is CD14-dependent [34, 117]. This activation may involve phosphorylation of a transcription factor by specific kinases (e.g. CREB/ATF1, or c-Jun, a member of the AP-1 family), or de novo synthesis as a part of LPS-induced 'early response genes' (e.g. JunB, c-Fos, or Egr-1). Activation of the NF-κB transcription factor is induced through TLR-2 [55, 56] and involves a pathway similar to the pathway induced by the IL-1 receptor. It includes adaptor proteins MyD88 and TRAF6, and kinases NIK, IKKα and IKKβ [56], which phosphorylate the NF-κB inhibitor, IκB. This phosphorylation induces degradation of IκB and release and translocation of NF-κB into the nucleus (fig. 2A).

Activated transcription factors bind to the specific sequences in the promoters of specific genes and induce transcription of these genes. Typically, activation of transcription of a given gene requires binding of several transcription factors. As shown in figure 2A, each transcription factor induced by LPS plays a role in regulation of transcription of several genes, and each gene is regulated by several transcription factors [34, 117–119].

Much less is known about transcription factors activated by PGN and other components of gram-positive bacteria. It is known that PGN induces CD14-dependent activation of NF-κB [8], most likely through TLR-2 [67a]. Recent results indicate that PGN also induces CD14-dependent activation of CREB/ATF1 and AP-1 [34]. PGN also activates Egr-1 and ATF2 [Gupta, unpubl.]. NF-κB, CREB/ATF1, and Egr-1 are all involved in the activation of transcription of TNF-α gene [Gupta, unpubl.] (fig. 2B). It is still not known which transcription factors are involved in the activation of other genes that are induced by PGN.

Acknowledgments

This work was supported by the National Institute of Health Grant AI28797 (to R.D. and D.G.), by a grant from the BMBF (No. 01KI9471) (to A.J.U.), and by the DFG (SFB 367, project C5) (to A.J.U.).

References

1   Wright SD, Ramos RA, Tobias PS, Ulevitch RJ, Mathison, JC: CD14, a receptor for complexes of lipopolysaccharide (LPS) and LPS binding protein. Science 1990;249:1431–1433.
2   Schumann RR, Leong SR, Flaggs GW, Gray PW, Wright SD, Mathison JC, Tobias PS, Ulevitch RJ: Structure and function of lipopolysaccharide binding protein. Science 1990;249:1429–1431.
3   Espevik T, Otterlei M, Skjåk-Bræk G, Ryan L, Wright SD, Sundan A: The involvement of CD14 in stimulation of cytokine production by uronic acid polymers. Eur J Immunol 1993;23:255–261.
4   Zhang Y, Doerfler M, Lee TC, Guillemin B, Rom WN: Mechanisms of stimulation of interleukin-1 beta and tumor necrosis factor-alpha by *Mycobycterium tuberculosis* components. J Clin Invest 1993;91:2076–2083.
5   Pugin J, Heumann D, Tomasz A, Kravchenko VV, Akamatsu Y, Nishijima M, Glauser MP, Tobias PS, Ulevitch RJ: CD14 is a pattern recognition receptor. Immunity 1994;1:509–516.
6   Weidemann B, Brade H, Rietschel ET, Dziarski R, Bazil V, Kusumoto S, Flad H-D, Ulmer AJ: Soluble peptidoglycan-induced monokine production can be blocked by anti-CD14 monoclonal antibodies and by lipid A partial structures. Infect Immun 1994;62:4709–4715.
7   Pedron T, Girard R, Turco SJ, Chaby R: Phosphatidylinositol-anchored molecules and inducible lipopolysaccharide binding sites of human and mouse bone marrow cells. J Biol Chem 1994;269:2426–2432.
8   Gupta D, Kirkland TN, Viriyakosol S, Dziarski R: CD14 is a cell-activating receptor for bacterial peptidoglycan. J Biol Chem 1996;271:23310–23316.
9   Weidemann B, Schletter J, Dziarski R, Kusumoto S, Stelter F, Rietschel ET, Flad HD, Ulmer AJ: Specific binding of soluble peptidoglycan and muramyldipeptide to CD14 on human monocytes. Infect Immun 1997;65:858–864.
10 Dziarski R, Tapping RI, Tobias P: Binding of bacterial peptidoglycan to CD14. J Biol Chem 1998;273:8680–8690.
11 Cleveland MG, Gorham JD, Murphy TL, Tuomanen E, Murphy KM: Lipoteichoic acid preparations of gram-positive bacteria induce interleukin-12 through a CD14-dependent pathway. Infect Immun 1996;64:1906–1912.
12 Kusunoki T, Hailman E, Juan TSC, Lichenstein HS, Wright SD: Molecules from *Staphylococcus aureus* that bind CD14 and stimulate innate immune responses. J Exp Med 1995;182:1673–1682.
13 Hattor Y, Kasai K, Akimoto K, Thiemermann C: Induction of NO synthesis by lipoteichoic acid from *Staphylococcus aureus* in J774 macrophages: Involvement of a CD14-dependent pathway. Biochem Biophys Res Commun 1997;233:375–379.
14 Soell M, Lett E, Holveck F, Scholler M, Wachsmann D, Klein JP: Activation of human monocytes by streptococcal rhamnose glucose polymers is mediated by CD14 antigen, and mannan binding protein inhibits TNF-alpha release. J Immunol 1995;154:851–860.

15 Cauwels A, Wan E, Leismann M, Tuomanen E: Coexistence of CD14-dependent and independent pathways for stimulation of human monocytes by gram-positive bacteria. Infect Immun 1997;65: 3255–3260.

16 Medvedev AE, Flo T, Ingalls RR, Golenbock DT, Teti G, Vogel SN, Espevik T: Involvement of CD14 and complement receptors CR3 and CR4 in nuclear factor-κB activation of TNF production induced by lipopolysaccharide and group B streptococcal cell walls. J Immunol 1998;160: 4535–4542.

17 Savedra R, Delude RL, Ingalls RR, Fenton MJ, Golenbock DT; Mycobacterial lipoarabinomannan recognition requires a receptor that shares components of the endotoxin signaling system. J Immunol 1996;157:2549–2554.

18 Sellati TJ, Bouis DA, Kitchens RL, Darveau RP, Pugin J, Ulevitch RJ, Gangloff SC, Goyert SM, Norgard MV, Radolf JD: *Treponema pallidum* and *Borrelia burgdorferi* lipoproteins and synthetic lipopeptides activate monocytic cells via a CD14-dependent pathway distinct from that used by lipopolysaccharide. J Immunol 1998;160:5455–5464.

19 Wooten M, Morrison TB, Weis JH, Wright SD, Thieringer R, Weis JJ: The role of CD14 in signaling mediated by outer membrane lipoproteins of *Borrelia burgdorferi*. J Immunol 1998;160:5485–5492.

20 Otterlei M, Varum KM, Ryan L, Espevik T: Characterization of binding and TNF-alpha-inducing ability of chitosans on monocytes: The involvement of CD14. Vaccine 1994;12:825–832.

21 Newman SL, Chaturvedi S, Klein BS: The WI-1 antigen of *Blastomyces dermatitidis* yeast mediates binding to human macrophage CD11/CD18 (CR3) and CD14. J Immunol 1995;154:753–761.

22 Cavaillon JM, Marie C, Caroff M, Ledur A, Godard I, Poulain D, Fitting C, Haeffner-Cavaillon N: CD14/LPS receptor exhibits lectin-like properties. J Endotoxin Res 1996;3:471–480.

23 Yu B, Hailman E, Wright SD: Lipopolysaccharide binding protein and soluble CD14 catalyze exchange of phospholipids. J Clin Invest 1997;99:315–324.

24 Wang PY, Kitchens RL, Munford RS: Phosphatidylinositides bind to plasma membrane CD14 and can prevent monocyte activation by bacterial lipopolysaccharide. J Biol Chem 1998;273:24309–24313.

25 Perera P-Y, Vogel SN, Detore GR, Haziot A, Goyert SM: CD14-dependent and CD14-independent signaling pathways in murine macrophages from normal and CD14 knockout mice stimulated with lipopolysaccharide and taxol. J Immunol 1997;158:4422–4429.

26 Bosco MC, Espinoza-Delgado I, Rowe TK, Malabarba MG, Longo DL, Varesio L: Functional role for the myeloid differentiation antigen CD14 in the activation of human monocytes by IL-2. J Immunol 1997;159:2922–2931.

27 Devitt A, Moffatt OD, Raykundalia C, Capra JD, Simmons DL, Gregory C: Human CD14 mediates recognition and phagocytosis of apoptotic cells. Nature 1998;392:505–509.

28 Schleifer KH, Kandler O: Peptidoglycan types in the bacterial cell walls and their taxonomic implications. Bact Rev 1975;36:407–477.

29 Rosenthal RS, Dziarski R: Isolation of peptidoglycan and soluble peptidoglycan fragments. Methods Enzymol 1994;235:253–285.

30 Heymer B, Seidl PH, Schleifer KH: Immunochemistry and biological activity of peptidoglycan; in Stewart-Tull DES, Davis M (eds): Immunology of the Bacterial Cell Envelope. New York, Wiley, 1985, pp 11–46.

31 Dziarski R: Effects of peptidoglycan on the cellular components of the immune system; in Seidl HP, Schleifer KH (eds): Biological Properties of Peptidoglycan. Berlin, Walter De Gruyter, 1986, pp 229–247.

32 Dziarski R, Ulmer AJ, Gupta D: Interactions of bacterial lipopolysaccharide and peptidoglycan with mammalian CD14; in Doyle RJ (ed): Glycomicrobiology. London, Plenum Press, 1999.

33 Mathison JC, Tobias PS, Wolfson E, Ulevitch RJ: Plasma lipopolysaccharide (LPS)-binding protein: A key component in macrophage recognition of gram-negative LPS. J Immunol 1992;149: 200–206.

34 Gupta D, Wang Q, Vinson C, Dziarski R: Bacterial peptidoglycan induces CD14-dependent activation of transcription factors CREB/ATF-1 and AP-1. J Biol Chem 1999;274:14012–14020.

35   Juan TS-C, Hailman E, Kelley MJ, Busse LA, Davy E, Empig CJ, Narhi LO, Wright SD, Lichenstein HS: Identification of a lipopolysaccharide binding domain in CD14 between amino acids 57 and 64. J Biol Chem 1995;270:5219–5224.

36   Stelter F, Bernheiden M, Menzel R, Jack RS, Witt S, Fan X, Pfister M, Schütt C: Mutation of amino acids 39–44 of human CD14 abrogates binding of lipopolysaccharide and *Escherichia coli*. Eur J Biochem 1997;243:100–109.

37   De Kimpe SJ, Kengatharan M, Thiemermann C, Vane JR: The cell wall components peptidoglycan and lipoteichoic acid from *Staphylococcus aureus* act in synergy to cause shock and multiple organ failure. Proc Natl Acad Sci USA 1995;92:10359–10363.

38   Kengatharan KM, De Kimpe S, Robson C, Foster SJ, Thiemermann C: Mechanism of gram-positive shock: Identification of peptidoglycan and lipoteichoic acid moieties essential in the induction of nitric oxide synthase, shock, and multiple organ failure. J Exp Med 1998;188:305–315.

39   Takada H, Kawabata Y, Arakaki R, Kusumoto S, Fukase K, Suda Y, Yoshimura T, Kokeguchi S, Kato K, Komuro T, Tanaka N, Saito M, Yoshida T, Sato M, Kotani S: Molecular and structural requirements of a lipoteichoic acid from *Enterococcus hirae* ATCC 9790 for cytokine-inducing, antitumor, and antigenic activities. Infect Immun 1995;63:57–65.

40   Suda Y, Tochio H, Kawano K, Takada H, Yoshida T, Kotani S, Kusumoto S: Cytokine-inducing glycolipids in the lipoteichoic acid fraction from *Enterococcus hirae* ATCC 9790. FEMS Immunol Med Microbiol 1995;12:97–112.

41   Bernardo J, Billingslea AM, Blumenthal RL, Seetoo KF, Simons ER, Fenton MJ: Differential responses of human mononuclear phagocytes to mycobacterial lipoarabinomannans: Role of CD14 and the mannose receptor. Infect Immun 1998;66:28–35.

42   Yu W, Soprana E, Costentino G, Volta M, Lichtenstein HS, Viale G, Vercelli D: Soluble CD141-152 confers responsiveness to both lipoarabinomannan and lipopolysaccharide in a novel HL-60 cell bioassay. J Immunol 1998;161:4244–4251.

43   Krziwon C, Zähringer U, Kawahara K, Weidemann B, Kusumoto S, Rietschel ETh, Flad HD, Ulmer AJ: Glycosphingolipids from *Sphingomonas paucimobilis* induce monokine production in human mononuclear cells. Infect Immun 1995;63:2899–2905.

44   Lee JD, Kato K, Tobias PS, Kirkland TN, Ulevitch RJ: Transfection of CD14 into 70Z/3 cells dramatically enhances the sensitivity to complexes of lipopolysaccharide (LPS) and LPS binding protein. J Exp Med 1992;175:1697–1705.

45   Pugin J, Schurer-Maly C-C, Leturq D, Moriarty A, Ulevitch RJ, Tobias PS: Lipopolysaccharide activation of human endothelial and epithelial cells is mediated by lipopolysaccharide-binding protein and soluble CD14. Proc Natl Acad Sci USA 1993;90:2744–2748.

46   Frey EA, Miller DS, Jahr TG, Sundan A, Bazil V, Espevik T, Finlay BB, Wright SD: Soluble CD14 participates in the response of cells to lipopolysaccharide. J Exp Med 1992;176:1665–1671.

47   Haziot A, Rong G-W, Silver J, Goyert SM: Recombinant soluble CD14 mediates the activation of endothelial cells by lipopolysaccharide. J Immunol 1993;151:1500–1507.

48   Read MA, Cordle SR, Veach RA, Carlisle CD, Hawiger J: Cell-free pool of CD14 mediates activation of transcription factor NF-κB by lipopolysaccharide in human endothelial cells. Proc Natl Acad Sci USA 1993;90:9887–9891.

49   Arditi M, Zhou J, Dorio R, Rong GW, Goyert SM, Kim KS: Endotoxin-mediated endothelial cell injury and activation: Role of soluble CD14. Infect Immun 1993;61:3149–3156.

50   Loppnow H, Stelter F, Schonbeck U, Schluter C, Ernst M, Schütt C, Flad H-D: Endotoxin activates human vascular smooth muscle endothelial cells despite lack of expression of CD14 mRNA or endogenous membrane CD14. Infect Immun 1995;63:1020–1026.

51   Orr SL, Tobias P: Endothelial and epithelial cell lines can be stimulated by mycobacterial lipoarabinomannan via membrane-bound CD14 but fail to respond via soluble CD14. 5th Conference of the International Endotoxin Society, Santa Fe, 1998, p 135.

52   Jin Y, Gupta D, Dziarski R: Endothelial and epithelial cells do not respond to complexes of peptidoglycan with soluble CD14, but are activated indirectly by peptidoglycan-induced tumor necrosis factor-α and interleukin-1 from monocytes. J Infect Dis 1998;177:1629–1638.

Interactions of CD14 with Components of Gram-Positive Bacteria   103

53   Golenbock DT, Liu Y, Millham FH, Freeman MW, Zoeller RA: Surface expression of human CD14 in Chinese hamster ovary fibroblasts imparts macrophage-like responsiveness to bacterial endotoxin. J Biol Chem 1993;268:22055–22059.
54   Delude RL, Savedra R, Zhao H, Thieringer R, Yamamoto S, Fenton MJ, Golenbock DT: CD14 enhances cellular responses to endotoxin without imparting ligand-specific recognition. Proc Natl Acad Sci USA 1995;92:9288–9292.
55   Yang R-B, Mark MR, Gray A, Huang A, Xie MH, Zhang M, Goddard A, Wood WI, Gurney AL, Godowski PJ: Toll-like receptor-2 mediates lipopolysaccharide-induced cellular signaling. Nature 1998;395:284–288.
56   Kirschning CJ, Wesche H, Ayres TM, Rothe M: Human Toll-like receptor 2 confers responsiveness to bacterial lipopolysaccharide. J Exp Med 1998;188:2091–2098.
57   Schumann RR, Pfeil D, Freyer D, Buerger W, Lamping N, Kirschning CJ, Goebel UB, Weber JR: Lipopolysaccharide and pneumococcal cell wall components activate the mitogen activated protein kinases (MAPK) erk-1, erk-2, and p38 in astrocytes. GLIA 1998;22:295–305.
58   Vita N, Lefort S, Sozzani P, Reeb R, Richards S, Borysiewicz L, Ferrara P, Labeta M: Detection and biochemical characteristics of the receptor for complexes of soluble CD14 and bacterial lipopolysaccharide. J Immunol 1997;158:3457–3462.
59   Tapping RI, Tobias PS: Cellular binding of soluble CD14 requires lipopolysaccharide (LPS) and LPS-binding protein. J Biol Chem 1997;272:23157–23164.
60   Bevilacqua MP: Endothelial-leukocyte adhesion molecules. Annu Rev Immunol 1993;11:767–804.
61   Wahl SM, Feldman GM, McCarthy JB: Regulation of leukocyte adhesion and signaling in inflammation and disease. J Leukoc Biol 1996;59:789–796.
62   Pugin J, Ulevitch RJ, Tobias PS: A critical role for monocytes and CD14 in endotoxin-induced endothelial cell activation. J Exp Med 1993;178:2193–2200.
63   Pugin J, Ulevitch RJ, Tobias PS: Tumor necrosis factor-α and interleukin-1β mediate human endothelial cell activation in blood at low endotoxin concentrations. J Inflamm 1995;45:49–55.
64   Wright SD: CD14 and innate recognition of bacteria. J Immunol 1995;155:6–8.
65   Ulmer AJ, Dziarski R, El-Samalouti V, Rietschel ET, Flad HD: CD14, an innate immune receptor for various bacterial cell wall components; in Morrison M (ed): Endotoxin in Health and Disease. New York, Marcel Dekker 1999, pp 463–472.
66   Ulevitch RJ, Tobias PS: Receptor-dependent mechanisms of cell stimulation by bacterial endotoxin. Annu Rev Immunol 1995;13:437–457.
67   Poltorak A, He X, Smirnova I, Liu M-Y, Van Huffel C, Du X, Birdwell D, Alejos E, Silva M, Galanos C, Freudenberg M, Riccardi-Castegnoli P, Layton B, Beutler B: Defective LPS signaling in C3H/HeJ and C57BL/10ScCr mice: Mutations in Tlr4 gene. Science 1998;282:2085–2088.
67a   Schwandner R, Dziarski R, Wesche H, Rothe M, Kirschning CJ: Peptidoglycan- and lipoteichoic acid-induced cell activation is mediated by Toll-like receptor 2. J Biol Chem 1999;274:17406–17409.
68   Harder T, Simons K: Caveolae, DIGs, and the dynamics of sphingolipid-cholesterol microdomains. Curr Opin Cell Biol 1997;9:534–542.
69   Brown DA, Rose JK: Sorting of GPI-anchored proteins to glycolipid-enriched membrane subdomains during transport to the apical cell surface. Cell 1992;68:533–544.
70   Cinek T, Horejsi V: The nature of large noncovalent complexes containing glycosyl-phosphatidyl-inositol-anchored membrane glycoproteins and protein tyrosine kinases. J Immunol 1992;149,2262–2270.
71   Stefanova I, Horejsi V, Ansotegui IJ, Knapp W, Stockinger H: GPI-anchored cell-surface molecules complexed to protein tyrosine kinases. Science 1991;254:1016–1019.
72   Stefanova I, Corcoran ML, Horak EM, Wahl LM, Bolen JB, Horak ID: Lipopolysaccharide induces activation of CD14-associated protein tyrosine kinase p53/56lyn. J Biol Chem 1993;268:20725–20728.
73   Wang PY, Kitchens RL, Munford RS: Bacterial lipopolysaccharide binds to CD14 in low-density domains of the monocyte-macrophage plasma membrane. J Inflamm 1995;47:126–137.
74   Pugin J, Kravchenko VV, Lee JD, Kline L, Ulevitch RJ, Tobias PS: Cell activation mediated by glycosylphosphatidylinositol-anchored or transmembrane forms of CD14. Infect Immun 1998;66:1174–1180.

75   Haziot A, Hijiya N, Schultz K, Zhang F, Gangloff SC, Goyert SM: CD14 plays no major role in shock induced by Staphylococcus aureus but down-regulates TNF-α production. J Immunol 1999; 162:4801–4805.

76   Kitchens RJ, Munford RS: CD14-dependent internalization of bacterial lipopolysaccharide (LPS) is strongly influenced by LPS aggregation but not by cellular responses to LPS. J Immunol 1998; 160:1920–1928.

77   Grunwald U, Fan X, Jack RS, Workalemahu G, Kallies A, Stelter F, Schütt C: Monocytes can phagocytose gram-negative bacteria by a CD14-dependent mechanism. J Immunol 1996;157:4119–4125.

78   Khanna KV, Choi CS, Gekker G, Peterson PK, Molitor TW: Differential infection of porcine alveolar macrophage subpopulations by nonopsonized Mycobacterium bovis involves CD14 receptors. J Leukoc Biol 1996;60:214–220.

79   Schiff DE, Kline L, Soldau K, Lee JD, Pugin J, Tobias PS, Ulevitch RJ: Phagocytosis of gram-negative bacteria by a unique CD14-dependent mechanism. J Leukocyt Biol 1997;62:786–794.

80   Muro M, Koseki T, Akifusa S, Kato S, Kowashi Y, Ohsaki Y, Yamamoto K, Nishijima M, Nishihara T: Role of CD14 molecules in internalization of Actinobacillus actinomycetemcomitans by macrophages and subsequent induction of apoptosis. Infect Immun 1997;65:1147–1151.

81   Gegner JA, Ulevitch RJ, Tobias PS: Lipopolysaccharide (LPS) signal transduction and clearance: Dual roles for LPS binding protein and membrane CD14. J Biol Chem 1995;270:5320–5325.

82   Kitchens RL, Munford RS: Enzymatically deacylated lipopolysaccharide (LPS) can antagonize LPS at multiple sites in the LPS recognition pathway. J Biol Chem 1995;270:9904–9910.

83   Takayama K, Mitchell DH, Din ZZ, Mukerjee P, Li C, Coleman DL: Monomeric Re lipopolysaccharide from Escherichia coli is more active than the aggregated form in the Limulus amebocyte lysate assay and in inducing Egr-1 mRNA in murine peritoneal macrophages. J Biol Chem 1994;269: 2241–2244.

84   Haziot A, Ferrero E, Kontgen F, Hijiya N, Yamamoto S, Silver J, Stewart CL, Goyert SM: Resistance to endotoxin shock and reduced dissemination of gram-negative bacteria in CD14-deficient mice. Immunity 1996;4:407–414.

85   Haziot A, Lin XY, Zhang F, Goyert SM: The induction of acute phase proteins by lipopolysaccharide uses a novel pathway that is CD14-independent. J Immunol 1998;160:2570–2572.

86   Ingalls RR, Golenbock DT: CD11c/CD18, a transmembrane signaling receptor for lipopolysaccharide. J Exp Med 1995;181:1473–1479.

87   Ingalls RR, Arnaout MA, Golenbock DT: Outside-in signaling by lipopolysaccharide through a tailless integrin. J Immunol 1997;159:433–438.

88   Ingalls RR, Monks BG, Savedra R, Christ WJ, Delude RL, Medvedev AE, Espevik T, Golenbock DT: CD11/CD18 and CD14 share a common lipid A signaling pathway. J Immunol 1998;161: 5413–5420.

89   Hamann L, El-Samalouti VT, Schletter J, Chyla I, Lentschat A, Flad HD, Rietschel ETh, Ulmer AJ: CD55, a new LPS-signaling element. 5th Conference of the International Endotoxin Society, Santa Fe, 1998, p 99.

90   Heumann D, Barras C, Severin A, Glauser MP, Tomasz A: Gram-positive cell walls stimulate synthesis of tumor necrosis factor alpha and interleukin-6 by human monocytes. Infect Immun 1994;62:2715–2721.

91   Gupta D, Jin Y, Dziarski R: Peptidoglycan induces transcription and secretion of TNF-α and activation of Lyn, extracellular signal-regulated kinase, and Rsk signal transduction proteins in mouse macrophages. J Immunol 1995;155:2620–2630.

92   Beaty CD, Franklin TL, Uehara Y, Wilson CB: Lipopolysaccharide-induced cytokine production in human monocytes: Role of tyrosine phosphorylation in transmembrane signal transduction. Eur J Immunol 1994;24:1278–1284.

93   English BK, Orlicek SL, Mei Z, Meals EA: Bacterial LPS and IFN-γ trigger the tyrosine phosphorylation of vav in macrophages: Evidence for involvement of the hck tyrosine kinase. J Leukocyt Biol 1997;62:859–864.

94   Crowley MT, Harmer SL, DeFranco AL: Activation-induced association of a 145-kDa tyrosine-phosphorylated protein with Shc and Syk in B lymphocytes and macrophages. J Biol Chem 1996; 271:1145–1152.

95 Crowley MT, Costello PS, Fitzer-Attas CJ, Turner M, Meng F, Lowell C, Tybulewicz VL, DeFranco AL: A critical role for Syk in signal transduction and phagocytosis by Fcγ receptors on macrophages. J Exp Med 1997;186:1027–1039.

96 Herrera-Velit P, Reiner NE: Bacterial lipopolysaccharide induces the association and coordinate activation of p53/56^{lyn} and phosphatidylinositol 3-kinase in human monocytes. J Immunol 1996; 156:1157–1165.

97 Herrera-Velit P, Knutson KL, Reiner NE: Phosphatidylinositol 3-kinase-dependent activation of protein kinase C-ζ in bacterial lipopolysaccharide-treated human monocytes. J Biol Chem 1997; 272:16445–16452.

98 Lowell CA, Berton G: Resistance to endotoxic shock and reduced neutrophil migration in mice deficient for the Src-family kinases Hck and Fgr. Proc Natl Acad Sci USA 1998;95:7580–7584.

99 Meng F, Lowell CA: Lipopolysaccharide (LPS)-induced macrophage activation and signal transduction in the absence of Src-family kinases Hck, Fgr, and Lyn. J Exp Med 1997;185:1661–1670.

100 Weinstein SL, Sanghera JS, Lemke K, DeFranco AL, Pelech SL: Bacterial lipopolysaccharide induces tyrosine phosphorylation and activation of mitogen-activated protein kinases in macrophages. J Biol Chem 1992;267:14955–14962.

101 Han J, Lee J-D, Tobias PS, Ulevitch RJ: Endotoxin induces rapid protein tyrosine phosphorylation in 70Z/3 cells expressing CD14. J Biol Chem 1993;268:25009–25014.

102 Han J, Lee J-D, Bibbs L, Ulevitch RJ: A MAP kinase targeted by endotoxin and hyperosmolarity in mammalian cells. Science 1994;265:808–811.

103 Liu MK, Herrera-Velit P, Brownsey RW, Reiner NE: CD14-dependent activation of protein kinase C and mitogen-activated protein kinases (p42 and p44) in human monocytes treated with bacterial lipopolysaccharide. J Exp Med 1994;153:2642–2652.

104 Raingeaud J, Gupta S, Rogers JS, Dickens M, Han J, Ulevitch RJ, Davis RJ: Proinflammatory cytokines and environmental stress cause p38 mitogen-activated protein kinase activation by dual phosphorylation on tyrosine and threonine. J Biol Chem 1995;270:7420–7426.

105 Dziarski R, Jin Y, Gupta D: Differential activation of extracellular signal-regulated kinase (ERK) 1, ERK2, p38, and c-Jun NH$_2$-terminal kinase mitogen-activated protein kinases by bacterial peptidoglycan. J Infect Dis 1996;174:777–785.

106 Hambelton J, Weinstein SL, Lem L, DeFranco AL: Activation of c-Jun N-terminal kinase in bacterial lipopolysaccharide-stimulated macrophages. Proc Natl Acad Sci USA 1996;93:2774–2778.

107 Geppert TD, Whitehurst CE, Thompson P, Beutler B: Lipopolysaccharide signals activation of tumor necrosis factor biosynthesis through the Ras/Raf-1/MEK/MAPK pathway. Mol Med 1994; 1:93–103.

108 Reimann T, Buscher D, Hipskind RA, Krautwald S, Lohmann-Matthes M-L, Bacarini M: Lipopolysaccharide induces activation of the Raf-1/MAP kinase pathway: A putative role for Raf-1 in the induction of the IL-1β and TNF-α genes. J Immunol 1994;153:5740–5749.

109 Buscher D, Hipskind RA, Krautwald S, Reimann T, Baccarini M: Ras-dependent and -independent pathways target the mitogen-activated protein kinase network in macrophages. Molec Cell Biol 1995;15:466–475.

110 Sanghera JS, Weinstein SL, Aluwalia M, Girn J, Pelech SL: Activation of multiple proline-directed kinases by bacterial lipopolysaccharide in murine macrophages. J Immunol 1996;156:4457–4465.

111 Swantek JL, Cobb MH, Geppert TD: Jun N-terminal kinase/stress-activated protein kinase (JNK/SAPK) is required for lipopolysaccharide stimulation of tumor necrosis factor-α (TNF-α) translation: Glucocorticoids inhibit TNF-α translation by blocking JNK/SAPK. Molec Cell Biol 1997;17: 6274–6282.

112 Yao Z, Diener K, Wang XS, Zukowski M, Matsumoto G, Zhou G, Mo R, Sasaki T, Nishina H, Hui CC, Tan TH, Woodgett JP, Penninger JM: Activation of stress-activated protein kinase/c-Jun N-terminal protein kinase (SAPKs/JNKs) by a novel mitogen-activated protein kinase kinase. J Biol Chem 1997;272:32378–32383.

113 Hambelton J, McMahon M, DeFranco AL: Activation of Raf-1 and mitogen-activated protein kinase in murine macrophages partially mimics lipopolysaccharide-induced signaling events. J Exp Med 1995;182:147–154.

114 Guthridge CJ, Eidlen D, Arend WP, Gutierrez-Hartmann A, Smith MF: Lipopolysaccharide and Raf-1 kinase regulate secretory interleukin-1 receptor antagonistic gene expression by mutually antagonistic mechanisms. Mol Cell Biol 1997;17:1118–1128.

115 Shapira L, Sylvia VL, Halabi A, Soskolne WA, Van Dyke TE, Dean DD, Boyan BD, Schwartz Z: Bacterial lipopolysaccharide induces early and late activation of protein kinase C in inflammatory macrophages by selective activation of PKC-ε. Biochem Biophys Res Commun 1997;240:629–634.

116 Novogrodsky A, Vanichkin A, Patya M, Gazit A, Osherov N, Levitzki A: Prevention of lipopolysaccharide-induced lethal toxicity by tyrosine kinase inhibitors. Science 1994;264:1319–1322.

117 Sweet MJ, Hume DA: Endotoxin signal transduction in macrophages. J Leukoc Biol 1996;60:8–26.

118 Groupp ER, Donovan-Peluso M: Lipopolysaccharide induction of THP-1 cells activates binding of c-Jun, Ets, and Egr-1 to the tissue factor promoter. J Biol Chem 1996;271:12423–12430.

119 Pan J, Xia L, Yao L, McEver R: Tumor necrosis factor-α or lipopolysaccharide-induced expression of the murine P-selectin gene in endothelial cells involves novel κB sites and a variant activating transcription factor/cAMP response element. J Biol Chem 1998;273:10068–10077.

Roman Dziarski, Northwest Center for Medical Education,
Indiana University School of Medicine, Gary, IN 46408 (USA)
Tel. +1 219 980 6535, Fax +1 219 980 6566, E-mail rdziar@iunhaw1.iun.indiana.edu

Jack RS (ed): CD14 in the Inflammatory Response.
Chem Immunol. Basel, Karger, 2000, vol 74, pp 108–121

....

Soluble CD14-Mediated Cellular Responses to Lipopolysaccharide

Richard I. Tapping, Peter S. Tobias

Department of Immunology, The Scripps Research Institute, La Jolla, Calif., USA

The outer monolayer of the outer membrane of gram-negative bacteria is comprised largely of an amphipathic glycolipid called lipopolysaccharide (LPS) or endotoxin. During infection, this component of gram-negative bacteria is a highly potent mediator of inflammatory responses. Upon activation with LPS, leukocytes in the blood release a multitude of inflammatory molecules that, although aimed at clearing the infection, often contribute to the development of a severe and often fatal medical condition known as sepsis [1, 2]. In addition to leukocytes, endothelial cells (ECs) also contribute significantly to the pathophysiologic events leading to gram-negative sepsis. In fact, the vascular endothelium itself actively participates in organ inflammation by expressing chemokines, cytokines and adhesion molecules.

Activation of Endothelial Cells by LPS

The ability of LPS to directly activate ECs has been recognized for almost two decades. Early studies revealed that human endothelial cells cultured in vitro display increased procoagulant activity, expression of tissue factor, and adhesiveness in response to LPS [3–5]. Subsequently, it was revealed that a variety of cell surface adhesion molecules are strongly expressed in response to LPS including ELAM-1 [6], ICAM-1 [7], and VCAM [8]. In addition, ECs were found to release the cytokines IL-1 [9–11] and IL-6 [12, 13] upon treatment with LPS. In contrast to the responses observed using human cells, a pronounced LPS-mediated cytotoxic effect was observed by researchers working with ECs from bovine sources [14]. In fact, LPS has a pronounced injurious effect on bovine EC morphology and metabolism leading to increased permeability of the endothelial layer [15].

CD14 was initially characterized as a cell-surface myeloid differentiation marker [16]. CD14 exists as both a glycosylphosphatidylinositol (GPI) anchored cell surface receptor and as a soluble molecule found in normal blood sera [17]. These two forms of CD14 are known as membrane CD14 (mCD14) or soluble CD14 (sCD14), respectively. As a glycoprotein CD14 carries both N- and O-linked carbohydrates [18]. Potentially, there are many sources for soluble CD14 including cleavage of the receptor from monocytes by proteases or phospholipases [19, 20], and direct secretion of full-length molecules which have bypassed the GPI-linking mechanism [21–23]. Monocytic stimulation-appears to result in the release of CD14 from the cell surface through the activation of a cellular protease [20].

In 1990, CD14 was identified as the receptor required for myeloid cells to respond to low concentrations of LPS [24]. Unfortunately, this finding did not explain the mechanism by which ECs are activated by LPS, since these cells were found to not express any measurable cell surface mCD14 [25]. Importantly, a number of studies revealed that the injurious effect of LPS on bovine endothelial cells appeared to require the presence of serum [15, 26]. Similarly, the LPS-dependent activation of human umbilical vein endothelial cells (HUVECs) was found to be much more pronounced when the cells were treated in the presence of human serum [27]. Another unexplained finding at the time was that a monoclonal antibody against CD14 completely blocked the cytotoxic effect of LPS on bovine endothelial cells [26]. These results were reconciled by reports, from a number of groups, that the factor in serum responsible for mediating responses of endothelial cells to LPS is in fact the soluble form of the CD14 receptor [28–34]. The necessity for sCD14 in these studies was demonstrated either by the loss of LPS responsiveness through immunodepletion of sCD14 from normal serum or by the acquisition of LPS responsiveness through the addition of recombinant sCD14 to serum-free media. Collectively, these papers demonstrate that responses of HUVECs to LPS, including the expression of ELAM-1, ICAM-1, and VCAM-1, the secretion of IL-6 and IL-8, as well as the activation of NF-kB, all require sCD14. Concurrently, the LPS-induced cytotoxicity of bovine endothelial cells was also found to require the soluble form of CD14 [28, 32, 34]. Thus, an exceptionally unusual feature of CD14 is its ability to mediate LPS-induced activation of cells as both a membrane receptor in one form and as a soluble cofactor in another.

Lipopolysaccharide-binding protein (LBP) is a molecule found in normal serum which enhances the ability of CD14-bearing macrophages to respond to LPS [35]. Biochemical and biophysical studies have revealed that LBP binds LPS and then delivers LPS to sCD14 thereby catalyzing the formation of sCD14-LPS complexes [36–38]. The activation of non-CD14-bearing cells by

LPS and sCD14 is enhanced by the presence of lipopolysaccharide-binding protein (LBP) [29]. Since the binding of LPS by sCD14 is a kinetically slow process in the absence of LBP, it is reasonable to hypothesize that the ability of LBP to enhance LPS and sCD14-dependent activation of ECs is due to its ability to catalyze the formation of sCD14-LPS complexes.

Soluble CD14-Mediated Cellular Activation

Interestingly, the ability of sCD14 to mediate LPS-induced cellular activation is not restricted to ECs. For example, the LPS-induced secretion of IL-8 from SW620 and HT29 cells, two colonic epithelial cell lines [29, 39], and IL-6 from U373 cells, an epithelial-like astrocytoma cell line [28], is mediated by sCD14. To date, an ever-growing list of cell types display sCD14-dependent responses to LPS including bronchial epithelial cells [40], tracheal epithelial cells [41], gingival fibroblasts [42], human vascular smooth muscle cells [43], and peripheral blood dendritic cells [44].

Several studies have revealed that the ability of sCD14 to mediate LPS-induced activation of cells is not restricted to nonmyeloid cells. Initially, sCD14 was found to enhance the LPS mediated release of IL-6 and TNF-α from monocytic cells [21, 45]. A direct effect of sCD14 on cellular responses of monocytic cells to LPS has been demonstrated most convincingly in macrophages derived from patients with paroxysmal nocturnal hemoglobinuria (PNH), in which hematopoietic cells fail to express GPI-linked proteins including CD14. PNH-derived macrophages devoid of mCD14 exhibit LPS-mediated activation which is dependent on the presence of sCD14 [46, 47]. Surprisingly, sCD14 also appears to at least enhance the LPS-mediated activation of CD14 bearing macrophages and neutrophils [46, 48]. Thus, CD14 clearly plays an important role as a soluble cofactor mediating activation of a variety of myeloid and nonmyeloid cell types in response to LPS.

CD14 knockout mice have been instrumental in assessing the role of CD14 in whole animals. As expected, these mice are highly resistant to shock induced either by LPS or live gram-negative bacteria [49]. In fact, LPS-mediated responses in CD14-deficient mice, including release of TNF and IL-6, requires the administration of much higher doses of LPS or whole bacteria compared to wild-type mice. In addition, peripheral blood mononuclear cells isolated from CD14-deficient mice do not exhibit release of TNF or IL-1β when they are exposed to LPS in the presence of sCD14-deficient autologous serum. However, the ability of CD14-deficient peripheral blood mononuclear cells to respond to LPS is partially restored through the addition of recombinant soluble CD14 to this assay [49]. These results confirm that myeloid cells can be activated by LPS through a sCD14-dependent pathway which acts independently of mCD14.

It is worthy to note that soluble forms of cell-surface receptors typically act to dampen the ligand mediated activation of cells by binding the ligand and making it unavailable to the receptor. In this regard, at high concentrations, sCD14 does act to dampen the responses of CD14-bearing cells to LPS. For example, recombinant sCD14 can inhibit the oxidative burst response of isolated human mononuclear cells [50]. In addition, LPS-induced TNF release in whole blood is inhibited by the addition of recombinant sCD14 [51]. Under certain conditions, sCD14 has also been shown to dampen the ability of LPS to prime neutrophils [52]. Moreover, mice are protected from LPS-induced lethality when they are injected with high doses of recombinant sCD14 [53, 54]. These findings have led to the proposal that sCD14 could be utilized directly as a therapeutic in the treatment of sepsis.

Taken together, with respect to mCD14-bearing cells, sCD14 appears to play a dual role as either an activator or inhibitor of LPS-mediated activation. The role that sCD14 assumes with respect to mCD14-bearing cells depends on the experimental conditions employed in the cell activation assay [52]. Although not completely understood at this time, there is evidence that sCD14-mediated cellular activation of mononuclear cells requires higher concentrations of LPS and, even under optimal conditions, is less robust than that mediated by mCD14 [49]. These findings suggest that LPS-induced signal transduction mediated by mCD14 is more efficient than that mediated by sCD14.

Structural Requirements of sCD14-Mediated Cellular Activation

The analysis of sCD14 mutants have revealed some unexpected structure function relationships. Most surprising is the finding that truncation of sCD14 at amino acid 152, a deletion removing more than half of the amino acids comprising sCD14, has no effect on the ability of the molecule to bind LPS or mediate activation of cells in the presence of LPS [55, 56]. This truncation removes 7 of the total of 10 leucine-rich repeats within CD14 revealing that the majority of these leucine-rich repeats are not required for the bioactivity of this molecule with respect to LPS. Other studies have revealed that amino acids 57–64 of sCD14 are protected from digestion with the protease Asp-N upon binding LPS [57]. Subsequent deletion mutagenesis of sCD14 confirmed a requirement for this region in binding LPS and in mediating cellular responses to LPS [58]. In support of this, an anti-CD14 antibody called MEM-18, which neutralizes the bioactivity of CD14, maps to the LPS-binding region comprising amino acids 57–64. However, this region of CD14 is not recognized by another neutralizing anti-CD14 antibody known as 3C10. Instead, the epitope for the 3C10 antibody maps close to the N-terminus of CD14 between amino acids 7 and 14 [59]. Indeed, the alanine replacement of amino acids 7–10 of sCD14 results in a molecule incapable of mediating LPS-induced cellular activation [59]

Structure function studies on CD14 have also been performed by expressing mutants of CD14 on the cell surface by transfection. Interestingly, amino acids 1–151 of CD14 fused to a heterologous GPI anchoring sequence results in a functional receptor mediating LPS-induced activation of CHO and 70Z/3 cells [60]. These investigators also found that deletion of amino acids 9–12, 22–25 or 59–63 of membrane CD14 results in diminished LPS binding and at least partial inhibition of LPS-mediated activation [61]. These results independently confirm that short regions in the vicinity of amino acids 10 and 60 of CD14 are essential for LPS binding and cellular activation. Moreover, it was found that deletion of amino acids 35–39 of CD14 abrogates LPS binding and cellular activation, defining another critical region for the bioactivity of CD14 [61]. A complete set of 23 alanine scanning mutants covering the N-terminal 152 amino acids of CD14 have been constructed and functionally tested by others [62]. Once again it was revealed that amino acids 9–13 are structurally associated with the LPS-binding site. The alanine replacement of amino acids 39–44 of CD14 also abrogates LPS-binding and cellular activation [62]. Although alanine replacement mutants cannot be directly compared to deletion mutants, both sets of results demonstrate that the region surrounding, or even simply comprising, amino acid 39 are essential for the LPS bioactivity of CD14. It is worthy to note that the epitope for most neutralizing anti-CD14 monoclonal antibodies map to both the region surrounding amino acid 10 and also the region surrounding amino acid 39 [61, 62]. Therefore, it is reasonable to predict that these two regions of CD14 lie within close proximity to each other and comprise a conformational epitope of this molecule. Obviously, these results could be put into better perspective with knowledge of the three-dimensional structure of CD14.

The fact that structure function analysis of CD14 reveals regions required for bioactivity which are similar for both the soluble and membrane anchored form would argue that these two forms of CD14 mediate LPS-induced cellular activation by a similar mechanism. In this regard, LPS-induced signal transduction of vascular endothelial cells appears similar to that of myeloid cells including the translocation of NF-kB and the phosphorylation of mitogen-activated protein kinases p38, ERK1 and ERK2 [63]. However, the analysis of identical CD14 deletion mutants, in the context of either a soluble protein or a membrane anchored receptor, reveals discrepancies between the ability of the two forms of CD14 to mediate cellular activation. For example, certain mutations that only partially inhibit LPS-mediated cellular activation of CD14 as a cell-surface receptor, completely ablate LPS-mediated activation of CD14 as a soluble protein [Kirkland, pers. commun.]. These results strongly suggest that mechanistic differences exist between cellular activation mediated by either soluble or membrane anchored forms of CD14. In support of this, two anti-CD14 monoclonal antibodies have been described that in the presence of LPS block sCD14-mediated activation of HUVECs, but have no effect on the responses of membrane CD14-bearing human monocytes to LPS [64].

In addition to binding and mediating responses to LPS, CD14 has many other biological and biochemical properties. In fact, CD14 mediates cellular responses to many pathogen-derived ligands including gram-positive peptido-glycan [65], mycobacterial lipoarabinomanan [66], fungal products [67, 68] and spirochetal lipoproteins [69, 70]. In this regard, a physical association between sCD14 and peptidoglycan has been demonstrated [71]. These findings establish CD14 as a versatile receptor and soluble cofactor of the innate immune system.

Mechanism of sCD14-Mediated Cellular Activation

The mechanism by which engagement of LPS by CD14 eventually leads to cellular activation is unknown. This statement is equally true in the case of both membrane and soluble CD14-mediated cellular activation as neither molecule traverses the plasma membrane. One theory as to how CD14 mediates cellular activation stems from its known biochemical actions. As mentioned previously, LBP is a serum molecule that acts to catalyze the transfer of LPS to sCD14 [36–38]. Surprisingly, LBP also appears to be able to mediate the transfer of a variety of phospholipids from phospholipid vesicles to sCD14 [72]. Recent studies have also shown that LBP and sCD14 are capable of transferring LPS to phospholipid vesicles [73, 74]. These findings have led to the proposal that the insertion of LPS into the phospholipid bilayer of a cell, either by membrane or soluble CD14, results in a packing geometry that perturbs the membrane and subsequently triggers cellular activation [75]. This hypothesis may be indirectly supported by the observed effects of certain pharmacologic agents on LPS-mediated signaling [75, 76]. Whether LBP and CD14 actually transfer LPS into the phospholipid bilayer of an intact cell, and whether this action would result in cellular activation, has yet to be established.

Another theory as to how CD14 mediates signaling is that it binds LPS and subsequently interacts with at least one as yet unidentified signal-transducing molecule [77, 78]. Thus, either membrane or soluble CD14 could be regarded as a protein subunit of a larger signal transducing complex which is activated by LPS. In this regard, it is known that lipid IVa, a precursor of LPS lacking two acyl chains, acts as an LPS agonist for human cells and as an LPS antagonist for mouse cells [79]. Although both these actions are mediated by CD14, the heterologous expression of mouse or human CD14 does not alter cellular responses to lipid IVa which remain species specific [80]. These findings demonstrate that the species specific effect must be mediated by a cell-signaling component, other than CD14, that effectively discriminates between LPS and lipid IVa [80]. To date, these results constitute the most compelling evidence for at least one accessory LPS-signaling molecule for CD14.

A variety of approaches have been taken to identify signaling partners for CD14. In the case of sCD14-mediated cellular activation, at least two groups have directly examined the physical association between radiolabeled sCD14 and whole cells [81, 82]. One group observed specific cell surface binding of radiolabeled sCD14 to HUVECs in the absence of LPS [81]. Another group observed specific and LPS-dependent association of radiolabeled sCD14 to several cell types [82]. In this later study, the cell association of labeled sCD14 required exceptionally high levels of LPS (5 μg/ml) and required at least 3 h to reach equilibrium at room temperature. Moreover, neither study, in both of which sCD14 was radiolabeled using ^{125}I and chloramine-T, demonstrated that the final radiolabeled preparation still mediated LPS-induced cell activation. In order to overcome the inactivation of sCD14 by chloramine-T which we have observed, we radiolabeled sCD14 to high specific activity (2,400 Ci/mmol) by engineering a protein kinase A site at the C-terminus [83]. Although the final radiolabeled sCD14 molecule retains the ability to fully mediate cellular responses to low concentrations of LPS, we were unable to observe either LPS-dependent or LPS-independent specific binding of sCD14 to any cell type [83]. These results indicate that any receptor for sCD14 is either of low cell-surface abundance or has a low affinity for sCD14 itself. A low affinity could indicate that sCD14 delivers LPS to another molecule and is not a physical part of any final complex leading to cell activation.

A family of receptors, collectively known as Toll receptors, are required for activation of the innate immune system in *Drosophila* [84, 85]. These receptors have a large N-terminal leucine-rich repeat region, a single transmembrane domain, and a C-terminal region with homology to that of the signaling domain of the IL-1 receptor. At least five of these Toll-like receptor (TLR) family members have been cloned from human libraries and named TLR1 through TLR5 [86, 87]. Consistent with their known role in *Drosophila*, a constitutively active form of human TLR4 was found to mediate innate immune responses in human cells [88]. The natural ligands for these receptors are not known, however, cotransfection of TLR2 and an NF-kB-driven reporter gene in 293T fibroblasts have implicated this receptor in mediating LPS signaling [89, 90]. At lower LPS concentrations the cellular activation observed is only moderately enhanced by mCD14, but appears to be largely dependent on the presence of purified sCD14 [90]. The idea that a TLR is involved in mediating LPS signaling has been supported by the finding that cotransfection of inactive genes for IL-1 signaling molecules inhibits NF-kB-driven reporter gene activity in both human dermal endothelial cells and in THP-1 monocytic cells [91]. Direct evidence for the role of a TLR in LPS-signaling stems from the recent finding that inbred mouse strains which are hyporesponsive to LPS, including the C3H/HeJ mouse and the C57BL/10ScCr mouse, have mutations in the gene encoding mouse TLR4 [92, 93]. That this truly represents the defect in these mice has been confirmed by the finding that targeted knockout of TLR4 results in mice that are hyporesponsive to LPS [94]. Collectively, the simplest interpretation of these findings is that certain TLRs represent signaling partners for membrane and soluble CD14. However, that any TLR is a direct receptor for LPS or LPS protein complexes has not been demonstrated. In fact, at this time the natural ligand for any of the mammalian TLRs remain to be discovered.

In serum obtained from healthy individuals the concentration of sCD14 is around 3–4 µg/ml [95]. A moderate, but statistically significant, increase in the serum levels of sCD14 is observed in sepsis patients [96, 97]. In one study, the average sCD14 concentration was 2.5 µg/ml in sera from healthy individuals and was 3.2 µg/ml in sera from patients with gram-negative septic shock [97]. Moreover, there was a statistically significant correlation between increased levels of sCD14 and high mortality [97]. This increased sCD14 may be mono-cyte derived as peripheral blood mononuclear cells isolated from septic patients have been shown to spontaneously secrete higher levels of sCD14 compared to those isolated from healthy individuals [98]. Interestingly, patients with periodontitis have significantly higher serum concentrations of sCD14 than healthy controls presumably due to chronic exposure to LPS [99].

In conclusion, the apparent biological function of sCD14 is that it allows non-mCD14-expressing cells, such as ECs, to respond to low levels of LPS. However, this statement is an oversimplification of the physiologic role of sCD14 during infection. For example, a variety of inflammatory mediators that are released from activated leukocytes during infection are also potent activators of ECs. So what is the purpose of the direct sCD14-mediated activation of ECs in response to LPS? It has been proposed that the direct activation may be required in order for ECs to attract neutrophils directly to the site of inflammation or may be required during the initial phase of infection [100]. On the other hand, it is clear that at high concentrations, sCD14 also acts to dampen immune responses to LPS. Although this inhibitory level of sCD14 is not achieved in blood during septic shock, it may be achieved in infected tissues where a high influx of neutrophils and macrophages, which appear to release sCD14, has taken place.

In the future it is anticipated that more fungal and bacterial products will be added to the list of natural ligands for CD14. A knowledge of the structure of CD14 may lead to an understanding of how this molecule physically interacts with such a seemingly diverse range of ligands. The mechanism by which CD14 mediates LPS-induced cell activation remains an important outstanding issue with respect to this remarkably versatile molecule of the innate immune system.

Acknowledgments

Supported by National Institute of Health Grants AI-32021 and HL-23584 to P.S.T. R.I.T. is a postdoctoral fellow of the American Heart Association Western States Affiliate. This is publication number 12401-IMM from The Scripps Research Institute.

References

1   Bone RC: Gram-negative sepsis. Background, clinical features, and intervention. Chest 1991;100: 802–808.
2   Glauser MP, Zanetti G, Baumgartner J-D, Cohen J: Septic shock: Pathogenesis. Lancet 1991;338: 732–736.
3   Colucci M, Balconi G, Lorenzet R, Pietra A, Locati D, Donati MB, Semeraro N: Cultured human endothelial cells generate tissue factor in response to endotoxin. Clin Invest 1983;71:1893–1896.
4   Schleimer RP, Rutledge BK: Cultured human vascular endothelial cells acquire adhesiveness for neutrophils after stimulation with interleukin 1, endotoxin, and tumor-promoting phorbol diesters. J Immunol 1986;136:649–654.
5   Pohlman TH, Stanness KA, Beatty PG, Ochs HD, Harlan JM: An endothelial cell surface factor(s) induced in vitro by lipopolysaccharide, interleukin 1, and tumor necrosis factor-alpha increases neutrophil adherence by a CDw18-dependent mechanism. J Immunol 1986;136:4548–4553.
6   Bevilacqua MP, Pober JS, Mendrick DL, Cotran RS, Gimbrone MA Jr: Identification of an inducible endothelial-leukocyte adhesion molecule. Proc Natl Acad Sci USA 1987;84:9238–9242.
7   Dustin ML, Springer TA: Lymphocyte function-associated antigen (LFA-1) interaction with intercellular adhesion molecule-1 (ICAM-1) is one of at least three mechanisms for lymphocyte adhesion to cultured endothelial cells. J Cell Biol 1988;107:321–331.
8   Carlos TM, Schwartz BR, Kovach NL, Yee E, Rosa M, Osborn L, Chi-Rosso G, Newman B, Lobb R: Vascular cell adhesion molecule-1 mediates lymphocyte adherence to cytokine-activated cultured human endothelial cells. Blood 1990;76[5]:965–970.
9   Stern DM, Bank I, Nawroth PP, Cassimeris J, Kisiel W, Fenton JW, Dinarello C, Chess L, Jaffe EA: Self-regulation of procoagulant events on the endothelial cell surface. J Exp Med 1985;162: 1223–1235.
10 Miossec P, Cavender D, Ziff M: Production of interleukin 1 by human endothelial cells. J Immunol 1986;136:2486–2491.
11 Libby P, Ordovas JM, Auger KR, Robbins AH, Birinyi LK, Dinarello CA: Endotoxin and tumor necrosis factor induce interleukin-1 gene expression in adult human vascular endothelial cells. Am J Pathol 1986;124:179–185.

12 Jirik FR, Podor TJ, Hirano T, Kishimoto T, Loskutoff DJ, Carson DA, Lotz M: Bacterial lipopolysaccharide and inflammatory mediators augment IL-6 secretion by human endothelial cells. J Immunol 1989;142:144–147.

13 Loppnow H, Libby P: Adult human vascular endothelial cells express the IL6 gene differentially in response to LPS or IL1. Cell Immunol 1989;122:493–503.

14 Harlan JM, Harker LA, Reidy MA, Gajdusek CM, Schwartz SM, Striker GE: Lipopolysaccharide-mediated bovine endothelial cell injury in vitro. Lab Invest 1983;48:269–274.

15 Meyrick BO, Ryan US, Brigham KL: Direct effects of *E. coli* endotoxin on structure and permeability of pulmonary endothelial monolayers and the endothelial layer of intimal explants. Am J Pathol 1986;122:140–151.

16 Bazil V, Horejsi V, Hilgert I, McMichael MJ (eds): Leucocyte Typing III. White Cell Differentiation Antigens. The Workshop: Myeloid Panel Antibodies Recognizing the 53-kDa Molecular Weight Monocyte Antigen (CD14). Oxford, Oxford University Press, 1982, p. 611–613.

17 Bazil V, Horejsi V, Baudys M, Kristofova H, Strominger J, Kostka W, Hilgert I: Biochemical characterization of a soluble form of the 53-kDa monocyte surface antigen. Eur J Immunol 1986; 16:1583–1589.

18 Stelter F, Pfister M, Bernheiden M, Jack RS, Bufler P, Engelmann H, Schutt C: The myeloid differentiation antigen CD14 is N- and O-glycosylated: Contribution of N-linked glycosylation to different soluble CD14 isoforms. Eur J Biochem 1996;236:457–464.

19 Bazil V, Baudys M, Hilgert I, Stefanova I, Low M, Brozek J, Horejsi V: Structural relationship between the soluble and membrane-bound forms of human monocyte surface glycoprotein CD14. Mol Immunol 1989;26:657–662.

20 Bazil V, Strominger JL: Shedding as a mechanism of down-modulation of CD14 on stimulated human monocytes. J Immunol 1991;147:1567–1574.

21 Labeta MO, Durieux J, Fernandez N, Herrmann R, Ferrara P: Release from a human monocyte-like cell line of two different soluble forms of the lipopolysaccharide receptor, CD14. Eur J Immunol 1993;23:2144–2151.

22 Durieux J-J, Vita N, Popescu O, Guette F, Calzada-Wack J, Munker R, Schmidt RE, Lupker J, Ferrara P, Ziegler-Heitbrock HWL, et al: The two soluble forms of the lipopolysaccharide receptor, CD14: Characterization and release by normal human monocytes. Eur J Immunol 1994;24:2006–2012.

23 Bufler P, Stiegler G, Schuchmann M, Hess S, Kruger C, Stelter F, Eckerskorn C, Schutt C, Engelmann H: Soluble lipopolysaccharide receptor (CD14) is released via two different mechanisms from human monocytes and CD14 transfectants. Eur J Immunol 1995;25:604–610.

24 Wright SD, Ramos RA, Tobias PS, Ulevitch RJ, Mathison JC: CD14, a receptor for complexes of lipopolysaccharide (LPS) and LPS binding proteins. Science 1990;249:1431–1433.

25 Beekhuizen H, Blokland I, Corsel-van Tilburg AJ, Koning F, van Furth R: CD14 contributes to the adherence of human monocytes to cytokine-stimulated endothelial cells. J Immunol 1991;147: 3761–3767.

26 Patrick D, Betts J, Frey EA, Prameya R, Dorovini-Zis K, Finlay BB: *Haemophilus influenzae* lipopolysaccharide disrupts confluent monolayers of bovine brain endothelial cells via a serum-dependent cytotoxic pathway. J Infect Dis 1992;165:865–872.

27 von Asmuth EJU, Leeuwenberg JFM, Ceska M, Buurman WA: LPS and cytokine-induced endothelial cell IL-6 release and ELAM-1 expression: Involvement of serum. Eur Cytokine Net 1991;2: 291–297.

28 Frey EA, Miller DS, Jahr TG, Sundan A, Bazil V, Espevik T, FInlay BB, Wright SD: Soluble CD14 participates in the response of cells to lipopolysaccharide. J Exp Med 1992;176:1665–1671.

29 Pugin J, Schurer-Maly CC, Leturcq D, Moriarty A, Ulevitch RJ, Tobias PS: Lipopolysaccharide (LPS) activation of human endothelial and epithelial cells is mediated by LPS binding protein and soluble CD14. Proc Natl Acad Sci USA 1993;90:2744–2748.

30 von Asmuth EJU, Dentener MA, Bazil V, Bouma MG, Leeuwenberg JFM, Buurman WA: Anti-CD14 antibodies reduce responses of cultured human endothelial cells to endotoxin. Immunology 1993;80:78–83.

31 Haziot A, Rong G-W, Silver J, Goyert SM: Recombinant soluble CD14 mediates the activation of endothelial cells by lipopolysaccharide. J Immunol 1993;151:1500–1507.

32   Arditi M, Zhou J, Dorio R, Rong GW, Goyert SM, Kim KS: Endotoxin-mediated endothelial cell injury and activation: Role of soluble CD14. Infect Immun 1993;61:3149–3156.

33   Read MA, Cordle SR, Veach RA, Carlisle CD, Hawiger J: Cell-free pool of CD14 mediates activation of transcription factor NF-kB by lipopolysaccharide in human endothelial cells. Proc Natl Acad Sci USA 1993;90:9887–9891.

34   Goldblum SE, Brann TW, Ding X, Pugin J, Tobias PS: Lipopolysaccharide (LPS)-binding protein and soluble CD14 function as accessory molecules for LPS-induced changes in endothelial barrier function, in vitro. J Clin Invest 1994;93:692–702.

35   Schumann RR, Leong SR, Flaggs GW, Gray PW, Wright SD, Mathison JC, Tobias PS, Ulevitch RJ: Structure and function of lipopolysaccharide binding protein. Science 1990;249:1429–1433.

36   Hailman E, Lichenstein HS, Wurfel MM, Miller DS, Johnson DA, Kelley M, Busse LA, Zukowski MM, Wright SD: Lipopolysaccharide (LPS)-binding protein accelerates the binding of LPS to CD14. J Exp Med 1994;179:269–277.

37   Tobias PS, Soldau K, Gegner JA, Mintz D, Ulevitch RJ: Lipopolysaccharide binding protein-mediated complexation of lipopolysaccharide with soluble CD14. J Biol Chem 1995;270:10482–10488.

38   Yu B, Wright SD: Catalytic properties of lipopolysaccharide (LPS) binding protein: transfer of LPS to sCD14. J Biol Chem 1996;271:4100–4105.

39   Schuerer-Maly C-C, Eckmann L, Kagnoff MF, Falco MT, Maly F-E: Colonic epithelial cells as a source of interleukin-8: Stimulation by inflammatory cytokines and bacterial lipopolysaccharide. Immunology 1994;81:85–91.

40   Striz I, Mio T, Adachi Y, Bazil V, Rennard S: The CD14 molecule participates in regulation of IL-8 and IL-6 release by bronchial epithelial cells. Immunology Letters 1998;62:177–181.

41   Diamond G, Russell JP, Bevins CL: Inducible expression of an antibiotic peptide gene in lipopoly-saccharide-challenged tracheal epithelial cells. Proc Natl Acad Sci USA 1996;93:5156–5160.

42   Hayashi J, Masaka T, Saito I, Ishikawa I: Soluble CD14 mediates lipopolysaccharide-induced intercellular adhesion molecule 1 expression in cultured human gingival fibroblasts. Infect Immun 1996;64:4946–4951.

43   Loppnow H, Stelter F, Schoenbeck U, Schluter C, Ernst M, Schutt C, Flad H-D: Endotoxin activates human vascular smooth muscle cells despite lack of expression of CD14 mRNA or endogenous membrane CD14. Infect Immun 1995;63:1020–1026.

44   Verhasselt V, Buelens C, Willems F, De Groote D, Haeffner-Cavaillon N, Goldman M: Bacterial lipopolysaccharide stimulates the production of cytokines and expression of costimulatory molecules by human peripheral blood dendritic cells: Evidence for a soluble CD14-dependent pathway. J Immunol 1997;158:2919–2925.

45   Sundan A, Ryan L, Brinch L, Espevik T, Waage A: The involvement of CD14 in stimulation of TNF production from peripheral mononuclear cells isolated from PNH patients. Scand J Immunol 1995;41:603–608.

46   Golenbock DT, Bach RR, Lichenstein H, Juan TS-C, Tadavarthy A, Moldow CF: Soluble CD14 promotes LPS activation of CD14-deficient PNH monocytes and endothelial cells. J Lab Clin Med 1995;125:662–671.

47   Schutt C, Schilling T, Grunwald U, Stelter F, Witt S, Kruger C, Jack RS: Human monocytes lacking the membrane-bound from of the bacterial lipopolysaccharide (LPS) receptor CD14 can mount an LPS-induced oxidative burst response mediated by a soluble form of CD14. Res Immunol 1995; 146:339–350.

48   Hailman E, Vasselon T, Kelley M, Busse LA, Hu MC-T, Lichenstein HS, Detmers PA, Wright SD: Stimulation of macrophages and neutophils by complexes of lipopolysaccharide and soluble CD14. J Immunol 1996;156:4384–4390.

49   Haziot A, Ferrero E, Kontgen F, Hijiya N, Yamamoto S, Silver J, Stewart CL, Goyert SM: Resistance to endotoxin shock and reduced dissemination of gram-negative bacteria in CD14-deficient mice. Immunity 1996;4:407–414.

50   Schutt C, Schilling T, Grunwald U, Schonfeld W, Kruger C: Endotoxin-neutralizing capacity of soluble CD14. Res Immunol 1992;143:71–78.

51   Haziot A, Rong G-W, Bazil V, Silver J, Goyert SM: Recombinant soluble CD14 inhibits LPS induced tumor necrosis factor-α produced by cells in whole blood. J Immunol 1994;152:5868–5876.

52 Troelstra A, Giepmans BNG, Van Kessel KPM, Lichenstein HS, Verhoef J, Van Strijp JAG: Dual effects of soluble CD14 on LPS priming of neutrophils. J Leukoc Biol 1997;61:173–178.

53 Haziot A, Rong GW, Lin X-L, Silver J, Goyert SM: Recombinant soluble CD14 prevents mortality in mice treated with endotoxin (lipopolysaccharde). J Immunol 1995;154:6529–6532.

54 Stelter F, Witt S, Furll B, Jack RS, Hartung T, Schutt C: Different efficacy of soluble CD14 treatment in high- and low-dose LPS models. Eur J Clin Invest 1998;28:205–213.

55 Juan TS-C, Kelley MJ, Johnson DA, Busse LA, Hailman E, Wright SD, Lichenstein HS: Soluble CD14 truncated at amino acid 152 binds lipopolysaccharide [LPS] and enables cellular response to LPS. J Biol Chem 1995;280:1382–1387.

56 Yu W, Soprana E, Cosentino G, Volta M, Lichenstein HS, Viale G, Vercelli D: Soluble CD14$_{1–152}$ confers responsiveness to both lipoarabinomannan and lipopolysaccharide in a novel HL-60 cell bioassay. J Immunol 1998;161:4244–4251.

57 McGinley MD, Narhi LO, Kelley MJ, Davy E, Robinson J, Rohde MF, Wright SD, Lichenstein HS: CD14: Physical properties and identification of an exposed site that is protected by lipopoly-saccharide. J Biol Chem 1995;270:5213–5218.

58 Juan TS-C, Hailman E, Kelley MJ, Busse LA, Davy E, Empig CJ, Narhi LO, Wright SD, Lichenstein HS: Identification of a lipopolysaccharide binding domain in CD14 between amino acids 57 and 64. J Biol Chem 1995;270:5219–5224.

59 Juan TSC, Hailman E, Kelley MJ, Wright SD, Lichenstein HS: Identification of a domain in soluble CD14 essential for lipopolysaccharide (LPS) signaling but not LPS binding. J Biol Chem 1995; 270:17237–17242.

60 Viriyakosol S, Kirkland TN: The N-terminal half of membrane CD14 is a functional cellular lipopolysaccharide receptor. Infect Immun 1996;64:653–656.

61 Viriyakosol S, Kirkland TN: A region of human CD14 required for lipopolysaccharide binding. J Biol Chem 1995;270:361–368.

62 Stelter F, Bernheiden M, Menzel R, Jack RS, Witt S, Fan X, Pfister M, Schutt C: Mutation of amino acids 39–44 of human CD14 abrogates binding of lipopolysaccharide and *Escherichia coli*. Eur J Biochem 1997;243:100–109.

63 Arditi M, Zhou J, Torres M, Durden DL, Stins M, Kim KS: Lipopolysaccharide stimulates the tyrosine phosphorylation of mitogen-activated protein kinases p44, p42, and p41 in vascular endo-thelial cells in a soluble CD14-dependent manner: Role of protein tyrosine in lipopolysaccharide-induced stimulation of endothelial cells. J Immunol 1995;155:3994–4003.

64 Haziot A, Katz I, Rong GW, Lin XY, Silver J, Goyert SM: Evidence that the receptor for soluble CD14:LPS complexes may not be the putative signal-transducing molecule associated with mem-brane-bound CD14. Scand J Immunol 1997;46:242–245.

65 Weidemann B, Brade H, Rietschel ET, Dziarski R, Bazil V, Kusumoto S, Flad H-D, Ulmer AJ: Soluble peptidoglycan-induced monokine production can be blocked by anti-CD14 monoclonal antibodies and by lipid A partial structures. Infect Immun 1994;62:4709–4715.

66 Zhang YM, Doerfler TC, Lee TC, Guillemin B, Rom WN: Mechanisms of stimulation of interleukin-1β and tumor necrosis factor-α by *Mycobacterium tuberculosis* components. J Clin Invest 1998;91: 2076–2083.

67 Espevik T, Otterlei M, Skjak-Braek G, Ryan L, Wright SD, Sundan A: The involvement of CD14 in stimulation of cytokine production by uronic acid polymers. Eur J Immunol 1993;23:255–261.

68 Pugin J, Heumann D, Tomasz A, Kravchenko V, Akamatsu Y, Nishijimi M, Glauser M-P, Tobias PS, Ulevitch RJ: CD14 is a pattern recognition receptor. Immunity 1994;1:509–516.

69 Selati TJ, Bouis DA, Kitchens RL, Darveau RP, Pugin J, Ulevitch RJ, Gangloff SC, Goyert SM, Norgard MV, Radolf JD: *Treponema pallidum* and *Borrelia burgdorferi* lipoproteins and synthetic lipopeptides activate monocytic cells via a CD14-dependent pathway distinct from that used by lipopolysaccharide. J Immunol 1998;160:5455–5464.

70 Wooten RM, Morrison TB, Weis JH, Wright SD, Thieringer R, Weis JJ: The role of CD14 in signaling mediated by outer membrane lipoproteins of *Borrelia burgdorferi*. J Immunol 1998;160: 5485–5492.

71 Dziarski R, Tapping RI, Tobias PS: Binding of bacterial peptidoglycan to CD14. J Biol Chem 1998;273:8680–8690.

72   Yu B, Hailman E, Wright SD: Lipopolysaccharide binding protein and soluble CD14 catalyze exchange of phospholipids. J Clin Invest 1997;99:315–324.

73   Wurfel MM, Hailman E, Wright SD: Soluble CD14 acts as a shuttle in the neutralization of lipopolysaccharide (LPS) by LPS-binding protein and reconstituted high density lipoprotein. J Exp Med 1995;181:1743–1754.

74   Wurfel MM, Wright SD: Lipopolysaccharide-binding protein and soluble CD14 transfer lipopoly-saccharide to phospholipid bilayers. J Immunol 1998;158:3925–3934.

75   Thieblemont N, Thieringer R, Wright SD: Innate immune recognition of bacterial lipopolysaccha-ride: Dependence on interactions with membrane lipids and endocytic movement. Immunity 1998; 8:771–777.

76   Detmers PA, Thieblemont N, Vasselon T, Pionkova R, Miller DS, Wright SD: Potential role of membrane internalization and vesicle fusion in adhesion of neutrophils in response to lipopolysaccha-ride and TNF. J Immunol 1996;157:5589–5596.

77   Ulevitch RJ, Tobias PS: Recognition of endotoxin by cells leading to transmembrane signalling. Curr Opin Immunol 1994;6:125–130.

78   Ulevitch RJ, Tobias PS: Receptor-dependent mechanisms of cell stimulation by bacterial endotoxin. Annu Rev Immunol 1995;13:437–457.

79   Golenbock DT, Hampton RY, Qureshi N, Takayama K, Raetz CRH: Lipid A-like molecules that antagonize the effects of endotoxins on human monocytes. J Biol Chem 1991;266:19490–19498.

80   Delude RL, Savedra R Jr, Zhao H, Thieringer R, Yamamoto S, Fenton MJ, Golenbock DT: CD14 enhances cellular responses to endotoxin without imparting ligand-specific recognition. Proc Natl Acad Sci USA 1995;92:9288–9292.

81   Zhang J-K, Morrison TK, Falk MC, Kang Y-H, Lee C-H: Characterization of the binding of soluble CD14 to human endothelial cells and mechanism of CD14-dependent cell activation by LPS. J Endotoxin Res 1996;3:307–315.

82   Vita N, Lefort S, Sozzani P, Reeb R, Richards S, Borysiewicz LK, Farrara P, Labeta MO: Dectection and biochemical characteristics of the receptor for complexes of soluble CD14 and bacterial lipopoly-saccharide. J Immunol 1997;158:3457–3462.

83   Tapping RI, Tobias PS: Cellular binding of soluble CD14 requires lipopolysaccharide (LPS) and LPS-binding protein. J Biol Chem 1997;272:23157–23164.

84   Lemaitre B, Nicolas E, Michaut L, Reichhart J-M, Hoffmann JA: The dorsoventral regulatory gene cassette spatzle/Toll/cactus controls the potent antifungal response in Drosophila adults. Cell 1996;86:973–983.

85   Williams MJ, Rodriguez A, Kimbrell DA, Eldon ED: The 18-wheeler mutation reveals complex antibacterial gene regulation in Drosophila host defense. EMBO J 1999;16:6120–6130.

86   Rock FL, Hardiman G, Timans JC, Kastelein RA, Bazan JF: A family of receptors structurally related to Drosophila Toll. Proc Natl Acad Sci USA 1998;95:588–593.

87   Chaudhary PM, Ferguson C, Nguyen V, Nguyen O, Massa HF, Eby M, Jasmin A, Trask BJ, Hood L, Nelson PS: Cloning and characterization of two Toll/interleukin-1 receptor-like genes TIL3 and TIL4: Evidence for a multigene receptor family in humans. Blood 1999;91:4020–4027.

88   Medzhitov R, Preston-Hurlburt P, Janeway CA: A human homologue of the Drosophila Toll protein signals activation of adaptive immunity. Nature 1998;388:394–397.

89   Yang R-B, Mark MR, Gray A, Huang A, Xie M-H, Zhang M, Goddard A, Wood WI, Gurney AL, Godowski PJ: Toll-like receptor-2 mediates lipopolysaccharide-induced cellular signalling. Nature 1998;395:284–288.

90   Kirschning CJ, Wesche H, Ayres TM, Rothe M: Human Toll-like receptor 2 confers responsiveness to bacterial lipopolysaccharide. J Exp Med 1999;188:2091–2097.

91   Zhang FX, Kirschning CJ, Mancinelli R, Xu X-P, Jin Y, Faure E, Mantovani A, Rothe M, Muzio M, Arditi M: Bacterial lipopolysaccharide activates nuclear factor-kB through interleukin-1 signaling mediators in cultured human dermal endothelial cells and human mononuclear phagocytes. J Biol Chem 1999;274:7611–7614.

92   Poltorak A, He.X., Smirnova I, Liu MY, Huffel CV, Birdwell D, Alejos E, Sivla M, Galanos C, Freudenberg M, et al: Defective LPS signaling in C3H/HeJ and C57BL/10ScCr mice: Mutations in the Tlr4 gene. Science 1998;282:2085–2088.

93 Qureshi ST, Lariviere L, Leveque G, Clermont S, Moore KJ, Gros P, Malo D: Endotoxin-tolerant mice have mutations in Toll-like receptor 4 (*Tlr4*). J Exp Med 1999;189:615–625.

94 Hoshino K, Takeuchi O, Kawai T, Sanjo H, Ogawa T, Takeda Y, Takeda K, Arika S: Toll-like receptor 4 (TLR4)-deficient mice are hyporesponsive to lipopolysaccharide: Evidence for TLR4 as the *lps* gene product. J Immunol 1999;162:3749–3752.

95 Grunwald U, Kruger C, Westermann J, Lukowsky A, Ehlers M, Schutt C: An enzyme linked immunoassay for the quantification of solubilized CD14 in biological fluids. J Immunol Methods 1992;155:225–232.

96 Kruger C, Schutt C, Obertacke U, Joka T, Muller FE, Knoller J, Koller M, Konig W, Schonfeld W: Serum CD14 levels in polytraumatized and severely burned patients. Clin Exp Immunol 1991; 85:297–301.

97 Landmann R, Zimmerli W, Sansano S, Link S, Hahn A, Glauser MP, Calandra T: Increased circulating soluble CD14 is associated with high mortality in gram-negative septic shock. J Infect Dis 1995;171:639–644.

98 Hiki N, Berger D, Prigl C, Boelke E, Wiedeck H, Seidelmann M, Staib L, Kaminishi M, Oohara T, Beger HG: Endotoxin binding and elimination by monocytes: Secretion of soluble CD14 represents an inducible mechanism counteracting reduced expression of membrane CD14 in patients with sepsis and in a patient with paroxysmal nocturnal hemoglobinuria. Infect Immun 1998;66:1135–1141.

99 Hayashi J, Masaka T, Ishikawa I: Increased levels of soluble CD14 in sera of periodontitis patients. Infect Immun 1999;67:417–420.

100 Pugin J, Ulevitch RJ, Tobias PS: A critical role for myeloid cells and CD14 in endotoxin-induced endothelial cell activation. J Exp Med 1993;178:2193–2200.

Peter S. Tobias, Department of Immunology, IMM-12,
The Scripps Research Institute, La Jolla, CA 92037 (USA)
Tel. +1 619 784 8215, Fax +1 619 784 8239, E-Mail tobias@scripps.edu

Jack RS (ed): CD14 in the Inflammatory Response.
Chem Immunol. Basel, Karger, 2000, vol 74, pp 122–140

....

Non-Inflammatory/Anti-Inflammatory CD14 Responses: CD14 in Apoptosis

Christopher D. Gregory

Institute of Cell Signalling and School of Biomedical Sciences, University of
Nottingham Medical School, Queen's Medical Centre, Nottingham, UK

Although the most renowned role of CD14 is that of a receptor for the
bacterial endotoxin, lipopolysaccharide (LPS) – with binding leading to pro-
inflammatory host responses that can contribute to septic shock syndrome –
it is becoming clear that CD14 can also elicit non-inflammatory or anti-
inflammatory responses. Significantly, CD14 has proved to function not only
as a receptor for components of potentially infectious agents, but also as
a molecule that is involved in the innate recognition of 'self' components.
Specifically, CD14 functions in the highly conserved and profoundly important
process of phagocytosis of cells engaged in physiological cell death, apoptosis
[1]. Out of necessity, apoptosis is non-phlogistic and the recognition and
phagocytic removal of 'apoptotic-self' by the innate immune system is rapid
and, in inflammatory terms, quiet. This chapter will consider the mechanisms
underlying the non-phlogistic nature of apoptosis with particular emphasis
on the role of CD14 in apoptotic-cell clearance.

Apoptosis: Cell Death without Inflammation

Physiological deletion of unwanted cells is a powerful means of cell-
population control in multicellular organisms. Apoptosis is the most widely
studied physiological cell-death process, having received massively increased
attention in recent years. It is profoundly important in the normal physiological cell deletion that occurs during embryogenesis and tissue modelling as
well as in adult tissue turnover. It is required to generate diversity in the
immune system and for the resolution of innate and adaptive immune re-

sponses. We now appreciate that dysfunctional apoptosis is also associated with a broad and increasing array of diseases including cancer, neurodegenerative disease, cardiovascular disease, chronic inflammation and autoimmune disease [2]. Identified largely on the basis of its morphological characteristics [3], apoptosis – a term used synonymously with 'programmed cell death' – results from the activation of a series of 'suicidal' biochemical steps that are orchestrated by the cell itself. In other words, the cell actively participates in its own demise.

The stereotypical morphological changes that occur in apoptosis include cell shrinkage, chromatin condensation and nuclear fragmentation [4]. Plasma membrane integrity appears to be maintained, at least during the early stages of the process, and the cell may undergo fragmentation into membrane-bound 'apoptotic bodies' which may either include, or exclude, nuclear material. Ultrastructurally, cytoplasmic organelles appear intact, although changes in mitochondrial morphology are sometimes prominent. The classical biochemical feature of apoptosis is cleavage of chromosomal DNA typically into oligonucleosome-sized fragments differing by 180–200 bp that are displayed as 'ladders' after electrophoresis of low-molecular-weight DNA [5]. Perhaps more commonly, higher order cleavage of DNA into 300- and 50-kb fragments also occurs [6]. The nuclease activity that is responsible, at least in part, for both high- and low-order DNA degradation observed during apoptosis has been identified recently as 'caspase activated DNase', CAD [7] (also known as DFF40 [8]). Indeed, the caspase family of cysteinyl proteases – currently consisting of at least 14 mammalian members – may provide key protease cascades that elicit and/or amplify the apoptosis program [9]. Caspase-mediated cleavage disables multiple intracellular proteins, including DNA repair enzymes and cytoskeletal components [10], but those that are critical for apoptosis have not been identified. Caspase cleavage that mediates *activation* of proteins involved in apoptosis provides a clearer picture. Thus, caspases, which are initially present in cells as inactive zymogens, are processed to active proteases by other caspase molecules [9]. This had led to the categorisation of individual enzymes as either 'initiator' or 'executioner' caspases. For example, caspase 8 may be regarded as an initiator caspase since it initiates downstream caspase (i.e. executioner caspase) activation after its recruitment to the cytoplasmic face of the plasma membrane following ligation of the death-receptor Fas (CD95) [11]. Other initiator caspases – for example caspase 9 – appear to be important in eliciting apoptosis as a result of mitochondrial membrane perturbation [12]. A well-studied executioner caspase, caspase 3, when activated by upstream caspases, results in activation of CAD. This is achieved through cleavage of a protein, ICAD (inhibitor of CAD, otherwise known as DFF45 [8, 13]), that is co-translationally expressed with CAD [7, 14]. ICAD therefore represents a further caspase substrate that leads to activation, rather than inhibition, of a biochemical process during apoptosis. Caspase activity is regulated by cellular and viral molecules, notably members of the Bcl-2 family of cytoplasmic apoptosis regulators [9, 15].

While the nature of the 'apoptosis engine' and the details of its regulation remain largely enigmatic, the characteristic of the apoptosis program that is arguably its most important is the 'flagging' of the cell surface that signals swift phagocytosis of the apoptotic cell. Deletion of individual cells by apoptosis occurs in multiple tissues as a critical homeostatic component. The phagocytic clearance process is so efficient that, even when apoptosis is occurring at high frequency, it is difficult to observe apoptotic cells or their remains in histological sections. Indeed this difficulty in appreciation of apoptosis in situ contributed significantly to the failure, until recently, of the biomedical research community at large to appreciate the profound importance of this form of cell death. By contrast, necrotic lesions – caused by accidental, passive cell death – are characterised by the death of continuous tracts of cells which are readily apparent histologically [4]. When observed, apoptotic cells and their remnants are most commonly encountered in the cytoplasm of phagocytes. Phagocytosis of apoptotic neighbours can be performed by various cell types which function as 'amateur' phagocytes (including fibroblasts [16], epithelial cells [17], Sertoli cells [18], smooth muscle cells [19] and mesangial cells [20]). The macrophages, however, are the professional phagocytes in the apoptotic-cell clearance process. Most importantly, apoptosis (in contrast to necrosis) occurs in the absence of inflammation [21]. Therefore, phagocytosis of apoptotic cells occurs (1) sufficiently rapidly so as to prevent histotoxic damage through leakage of intracellular components from the dying cells, and (2) in a manner that fails to activate pro-inflammatory responses in the phagocytes. For the professional phagocyte, the quiet engulfment of apoptotic cells contrasts markedly with the pro-inflammatory responses that are elicited upon engagement, for example, of its Fcγ receptors [22, 23]. As will be discussed below, macrophages mediating clearance of apoptotic cells may deploy, on the one hand, receptors that are distinct from those mediating clearance of microbes and their products (pro-inflammatory clearance), and on the other, receptors that are common to clearance both of 'safe' (apoptotic self) and of dangerous (e.g. infectious) components. CD14 is a receptor which falls into the latter category.

Accumulating evidence indicates that persistence of apoptotic cells through failed phagocytic clearance is likely to contribute significantly to the pathogenesis of auto-immune and chronic inflammatory diseases [21–27]. Intuitively, the uncleared apoptotic cell in vivo might be expected, just as the apoptotic cell in vitro that persists in the absence of phagocytes, to take on some of the characteristics of necrotic cells: in particular, loss of plasma membrane integrity and consequent externalisation of intracellular macromolecules which may cause either direct damage to neighbouring cells or indirect damage by signalling for inflammatory responses or by acting as a source of auto-antigens.

The Apoptotic-Cell Surface

The first investigations into the properties of the apoptotic-cell surface that support recognition by phagocytes were made by Wyllie and colleagues in the early 1980s [28, 29]. Working on the adhesion of apoptotic rodent thymocytes to peritoneal macrophages that occurs at low temperature, they showed that the surface of the apoptotic cell changed rapidly following induction of apoptosis, allowing swift macrophage binding to occur in the absence of serum factors. Morphologically, the apoptotic-cell surface was observed to lose microvilli – leading to a reduction in surface area – and displayed a pitted structure, possibly resulting from the fusion of endoplasmic reticulum-derived and Golgi-derived vesicles with the cell surface. Apoptotic thymocytes were also found to have a reduced surface charge and were inhibited in their ability to interact with macrophages by certain sugars, particularly N-acetyl glucosamine, and N,N′-diacetyl chitobiose, but not by others, notably *D*-mannose and N-acetyl neuraminic acid. These observations, together with more recent studies that have demonstrated inhibition by sugars of apoptotic-cell interactions with both amateur and professional phagocytes [16, 30–32], suggest that changes in the sugar composition of glycoproteins and glycolipids on the apoptotic-cell surface may permit interaction with phagocyte receptors possessing lectin-like properties. As will be discussed below, macrophage CD14 is likely to function as such a receptor.

Phosphatidylserine (PS), the major anionic phospholipid of the plasma membrane is normally restricted to the inner membrane leaflet. During apoptosis, plasma membrane phospholipid asymmetry is lost and PS 'flips' to the outer membrane leaflet (fig. 1). This is probably the most widely known characteristic of the apoptotic-cell surface, since the Ca^{2+}-dependent binding of annexin V to externalised PS (on cells that have intact plasma membranes) is a commonly used indicator of apoptosis [33]. The mechanistic basis for the externalisation of PS during apoptosis has not yet been worked out in detail but appears to involve a non-specific lipid 'scramblase' which is activated during apoptosis together with an ATP-dependent aminophospholipid translocase, which is inhibited in apoptosis [34, 35]. It seems that the scramblase is important in mediating the rapid externalisation of PS during apoptosis, while

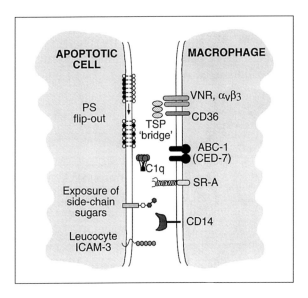

Fig. 1. 'Close encounters' between apoptotic cells and macrophages. See text for details. Note that no receptor/ligand interactions have yet been defined. ABC = ATP-binding cassette; SR-A = class A scavenger receptor; TSP = thrombospondin; VNR = vitronectin receptor.

the aminophospholipid translocase probably determines the absolute level of PS that persists on the outside of the apoptotic cell [36]. That PS on apoptotic cells can function in mediating apoptotic-cell clearance by macrophages was originally demonstrated using PS-containing liposomes which were found to inhibit markedly the uptake of apoptotic leukocytes by certain macrophages, notably murine thioglycollate-elicited macrophages which preferentially remove apoptotic cells in vitro by PS-dependent mechanisms, presumably through activity of one or more types of PS receptor [37, 38]. Exposure of PS may precede the nuclear changes that are characteristic of apoptosis but it is not yet clear how or if the pieces of the apoptosis machinery thus far identified couple with the mechanism of PS 'flip-out'. Thus, anti-apoptotic Bcl-2 expression and caspase inhibition prevent PS externalisation [39, 40]. One possible contributory factor to loss of PS asymmetry may be related to the loss of cytoskeletal anchorage of PS to fodrin which is cleaved by caspase activity during apoptosis [41]. However, it should be noted that caspase-3 activation and other consequent features of apoptosis can be uncoupled from PS externalisation [42], indicating that present knowledge of the apoptosis machinery does not extend to its effects on the cell surface.

Additional, ill-defined, features of apoptotic-cell surfaces include thrombospondin 1-binding sites, C1q-binding sites and, on apoptotic leukocytes, qualitative changes in the immunoglobulin superfamily member, intercellular adhesion molecule (ICAM)-3 (fig. 1). The existence of thrombospondin 1-binding sites is implied from studies of the phagocytosis of apoptotic leukocytes that is dependent upon a macrophage receptor complex composed of the $\alpha v \beta 3$ integrin and the class B scavenger receptor CD36 (see below). $\alpha v \beta 3$/CD36-dependent apoptotic-cell phagocytosis was found to be mediated by thrombospondin 1, which provides a molecular 'bridge' between the phagocyte receptor complex and the apoptotic-cell surface [43]. It seems possible that C1q performs a similar function since C1q-knockout mice display a defect in apoptotic-cell clearance capacity that is manifest in glomerulonephritis caused by persistence of apoptotic neutrophils [26]. The potential involvement of complement in apoptotic-cell clearance is intriguing since although complement components may serve, at least under certain circumstances, to enhance phagocytosis of apoptotic cells [26, 44], the clearance process can occur (in vitro) effectively in the absence of complement.

Work from this laboratory has demonstrated a role in the clearance of apoptotic leukocytes for ICAM-3, an adhesion molecule that is expressed by, and restricted to, the majority of leukocytes. We found that ICAM-3 participates in the removal of apoptotic leukocytes by gaining capacity to interact with macrophage receptor(s) that engage in apoptotic-cell clearance through binding to a site on the most membrane distal of the five immunoglobulin-like domains of ICAM-3 [45]. This site appears to be close to or coincident with the site of ICAM-3 that binds to its best-known receptor, the leukointegrin, LFA-1. However, the apoptotic form of ICAM-3 appears to be incapable of interacting with LFA-1, indicating that, on leukocytes engaged in apoptosis, ICAM-3 undergoes a qualitative change that results in loss of ability to bind its 'viable-state' receptor, LFA-1, and gain in ability to bind its 'apoptotic-state' receptor(s). Preliminary evidence suggests that one of these receptors may be CD14 [45]. This receptor switching of ICAM-3 might occur as part of the apoptotic leukocyte's tendency to isolate itself from its environment. Thus, ICAM-3/LFA-1 interactions are important in eliciting primary immune responses [46–50]; the switch to ICAM-3/CD14 interactions might be expected, for example, to militate against abortive or aberrant immune responses through ICAM-3 on pro-apoptotic antigen-presenting cells interacting with LFA-1 on T lymphocytes.

Macrophage Receptors Implicated in Apoptotic-Cell Clearance

The multiplicity of cell-surface changes that occur during apoptosis is reflected in the growing number of phagocyte receptors that are becoming implicated in supporting the clearance process (fig. 1). Amongst macrophage receptors, clearest evidence for functional activity in apoptotic-cell clearance has been presented for the αvβ3 integrin (vitronectin receptor) [51], CD36 [43], ATP-binding cassette transporter, ABC-1 [52], class A scavenger receptor [53], and CD14 [1]. Despite this panoply of receptors, no definitive phagocyte-receptor/apoptotic-cell-ligand pairing has yet been demonstrated. Significantly, none of these receptors appears to interact preferentially with PS to mediate engulfment of apoptotic cells, suggesting that additional receptors have yet to be defined [36].

Why do so many potential receptor-ligand interactions exist? One possibility is that multiplicity reflects redundancy, highlighting the importance of the clearance process and the requirement for failsafe functional activity, since persistence of apoptotic cells, as discussed above, is potentially pathogenic. Another is that multiple pathways represent first-line and back-up mechanisms supporting clearance of cells in different phases of apoptosis [24]. During high-rate apoptosis, for example at sites of inflammation where apoptotic-cell output exceeds phagocyte input, back-up mechanisms might be expected to be particularly important. Finally, different receptor-ligand interactions may predominate according to the lineage or phenotype of the phagocyte. For example, some macrophages preferentially utilise PS receptors in engulfing apoptotic cells, others preferentially use αvβ3/CD36 or CD14 [38, 54, 55]. Dendritic cells appear to phagocytose apoptotic cells via αvβ5 and CD36 rather than via CD14 [56, 57] indicating that specific receptors may support engulfment with specific immunological consequences. Since dendritic cells are capable of presenting antigens from apoptotic cells they have engulfed, and subsequently inciting immune responses to those antigens [56], one may speculate that CD14-dependent engulfment may be tolerogenic while certain alternative pathways, such as αvβ5/CD36, are immunogenic. The apparent complexity in receptor-ligand interactions that supports apoptotic-cell clearance in mammals may not be altogether surprising when one considers evolutionary aspects of the process. Thus, in *Caenorhabditis elegans*, a nematode worm, in which most of the genetic studies of developmental cell death have been performed, at least six genes (*ced*-1, -2, -5, -6, -7 and -10) have thus far been implicated in the clearance process [58], the product of one (*ced*-7) being a worm homologue of mammalian ABC-1. So complexity in clearance is already apparent in an invertebrate that, incidentally, does not possess a macrophage equivalent.

Interactions between CD14 and Apoptotic Cells

The first indication that CD14 can function in the clearance of apoptotic cells by macrophages emerged from our original observations that a mono-clonal antibody (mAb), 61D3, could inhibit the recognition and phagocytosis of apoptotic leukocytes by human monocyte-derived macrophages through binding to an undefined epitope on the macrophage surface [54]. Transient expression cloning in COS cells probed with 61D3 led us to the cDNA encoding the 61D3-reactive protein. This proved to be CD14 [1]. This observation linked a macrophage receptor renowned for its involvement in innate immune pro-inflammatory responses with the non-phlogistic removal of 'unwanted self'. Furthermore, those CD14 mAbs that inhibited interaction of apoptotic cells with macrophages mapped to a common epitope on CD14 that was bound by MEM18, a mAb that blocks LPS-binding to CD14, indicating that, if macrophage CD14 binds directly to apoptotic cells, the apoptotic-cell-associ-ated ligand(s) impinge on the LPS-binding site of CD14 [1]. Yet, as will be discusssed below, the macrophage response following apoptic-cell binding to CD14 differs profoundly from that consequent to LPS binding.

The mechanisms by which CD14 mediates clearance of apoptotic cells have not been defined but it seems reasonable (at least until experimental evidence indicates otherwise) to regard macrophage CD14 as a membrane receptor that interacts directly with one or more classes of apoptotic cell-associated ligand. Present evidence supports this supposition: COS cells over-expressing surface CD14 are enhanced in their capacity to bind and phagocy-tose apoptotic cells [1]. In addition, CD14-dependent apoptotic-cell clearance can occur in the absence of serum [1, 54]. However, the possibility that soluble CD14 can contribute to apoptotic-cell clearance by 'bridging' apoptotic-cell-associated surface components with macrophage receptors other than CD14 – in a manner analogous to that proposed for thrombospondin above – has not yet been addressed. Macrophage molecules that interact with such CD14/apoptotic-cell complexes may be identical or related to those described recently [59–61]. Again taking the thrombospondin 'bridge' as a precedent mechanism by which macrophages can interact with apoptotic cells, it also remains possible that CD14, as a macrophage receptor, binds apoptotic-cell surface components indirectly through bridging intermediaries (see below). Of course, these mecha-nisms may not be mutually exclusive (fig. 2).

The simple notion that GPI-linked CD14 acts as a receptor for apoptotic-cell surface structures poses two straightforward questions: (1) What is the identity of the apoptotic-cell-associated ligand(s)? and (2) How does ligand-binding to such a GPI-linked receptor elicit the subsequent macrophage re-sponse? These are addressed below.

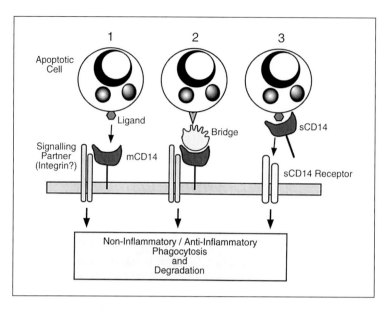

Fig. 2. Modes of interaction between CD14 and apoptotic cells that could support apoptotic-cell clearance by macrophages. 1 = Simple direct ligand/receptor interaction. 2 = Involvement of intermediate molecular bridge. 3 = Involvement of soluble CD14 (sCD14). See text for details. mCD14 = GPI-linked membrane CD14.

Ligands

There is some evidence that the propensity for clearance by macrophage CD14 is a characteristic of the lineage of the apoptotic cell, with apoptotic-lymphocyte clearance being more CD14-dependent than that of apoptotic neutrophils [54, 55]. Our recent studies using multiple lineages of apoptotic cells, however, indicate that human monocyte-derived macrophages engage, to a greater or lesser degree, in CD14-dependent clearance of all apoptotic cell types tested, regardless of the apoptosis stimulus [Devitt et al., unpubl.]. It would appear therefore that apoptotic-cell-associated CD14 ligands are ubiquitous. Although LPS remains the best known class of CD14 ligands, numerous additional ligands of various classes, including phospholipids, lipoproteins and polysaccharides, have been reported [62]. Two properties of CD14 may be relevant to its ability to interact with apoptotic-cell-associated ligands: one is its capacity to bind phospholipids [63, 64], the other is its lectin-like activity [65–67].

Perhaps the most obvious apoptotic-cell-surface component with potential to interact with macrophage CD14 is PS. CD14 has proven ability to bind to phosphatidylcholine, phosphatidylinositol (PI), phosphatidylethanolamine and PS, of which only PS has been clearly implicated in apoptotic-cell clearance – although as yet not in the context of a CD14-mediated route. However, a recent study demonstrated that PI binds to CD14 with higher affinity than PS [64]. Notably, PI appears to bind to the same site on GPI-linked CD14 as LPS and can modulate responses to LPS [64]. Where tested, however, PI-containing liposomes, unlike PS-containing liposomes, failed to inhibit apoptotic-cell clearance [37]. In recent studies, we have found no evidence that CD14, either cell-bound or soluble, preferentially binds PS. We can also clearly demonstrate CD14-mediated clearance of apoptotic cells by human monocyte-derived macrophages that is resistant to inhibition both by PS-containing liposomes and by lipid-symmetric erythrocytes [Devitt et al., in prep.]. The conclusion from this work is that PS is not a preferential apoptotic cell-associated ligand for CD14 on human monocyte-derived macrophages. This conclusion does not imply that human macrophage CD14 fails to bind apoptotic-cell-associated PS under all circumstances. Interaction of apoptotic cells with β-glucan-treated human monocyte-derived macrophages has a CD14-mediated component *and* a PS-mediated component [55; Devitt et al., in prep.]. The extent to which these components may overlap in a receptor-ligand pairing has not been investigated. In addition, if soluble CD14 proves to provide an alternative CD14-mediated apoptotic-cell clearance pathway, it will be important to determine the role of PS in fixing CD14 to the apoptotic-cell surface (fig. 2). While the context of PS (liposome versus apoptotic cell) might be an additional consideration that might affect its ability to interact with CD14, it seems probable that CD14 is not the major macrophage receptor for apoptotic-cell-associated PS. The identity of the macrophage PS-receptor(s) that are functional in apoptotic-cell clearance are eagerly awaited.

How else might macrophage CD14 interact with apoptotic cells? One possibility is by indirect means through a molecular bridge which binds to externalised phospholipids on the lipid-symmetric apoptotic-cell surface. Such a bridging function has been proposed for β(2)-glycoprotein I. However, even if this complex proves to be functional in apoptotic-cell clearance, it does not appear to interact with CD14 [68]. Another potential ligand for CD14 is the intercellular adhesion molecule, ICAM-3 (CD50). As already noted above, the surface of leukocytes becomes altered during apoptosis such that ICAM-3, a highly glycosylated member of the immunoglobulin superfamily, switches in receptor-binding preference from LFA-1, to which the 'viable form' of ICAM-3 binds, to one or more macrophage receptors involved in apoptotic-cell clearance [45]. Preliminary evidence points towards macrophage CD14 being involved in ICAM-3-dependent apoptotic-leukocyte clearance. This evidence is based on the observation that overexpression of ICAM-3 in human embryonic kidney cells (HEK), which do not normally express this leukocyte-restricted molecule, leads to enhancement in the ability of the HEK/ICAM-3 cells to interact with macrophages following induction of apoptosis. The enhanced macrophage interaction can be shown to be ICAM-3-dependent by inhibitory ICAM-3 mAbs [45]. Inhibitory CD14 mAbs, such as 61D3, also block this enhanced ability of apoptotic HEK/ICAM-3 cells to interact with macrophages whereas an inhibitory αv-specific mAb 13C2 fails to affect the interaction. Significantly, the combination of inhibitory ICAM-3 and CD14 mAbs has no additive effect in blocking the interaction, lending support to the view that ICAM-3 interacts with a macrophage clearance pathway that involves CD14 [45]. The simplest view is that CD14 is a macrophage receptor for 'apoptotic' ICAM-3 but direct interaction between ICAM-3 and CD14 has yet to be demonstrated. If such an interaction proves to be the case, it will be important to determine its molecular basis. It is tempting to speculate that, given that ICAM-3 is a massively glycosylated protein [69], CD14/ICAM-3 interactions might be lectin/sugar-mediated. Whether or not ICAM-3 proves to be a true, direct ligand on apoptotic leukocytes for CD14, it is clear from the studies so far that it is not the sole apoptotic-cell-associated ligand for this receptor – since apoptotic nonleukocytes (which never naturally express ICAM-3) can also be phagocytosed by CD14-dependent means [45]. Again, one can only speculate at this point for want of further experimental evidence but it seems likely that multiple apoptotic-cell-surface components can engage with the CD14-dependent clearance pathway. The identification of the key molecules on apoptotic cells that are involved in CD14-dependent clearance by macrophages and of the molecular basis of their interaction with CD14, direct or indirect, are important areas for future investigation.

Signalling Mechanisms: The Apoptotic Cell as an Anti-Inflammatory Signalling Device

CD14's role as a pattern-recognition receptor [70] accords with its potential to function as a receptor for 'apoptotic self'. For reasons already discussed, it seems likely that apoptotic cells can interact directly with GPI-anchored cell-surface CD14. One recent suggestion [71] which is consistent with the present data, is that CD14 acts as a 'tethering' molecule for apoptotic cells which subsequently interacts with other macrophage receptors involved in apoptotic-cell clearance. This proposal addresses the question of how the GPI-linked CD14 transduces intracellular signals following ligand binding. This question is a contentious one since, following LPS-binding to CD14, there is evidence supporting signalling via (1) the activity of transmembrane 'CD14-

partner' proteins that function as promiscuous signal-transducing components of a CD14 receptor complex [72, 73], and (2) lipid transfer [74, 75; see paper by Kitchens, this vol.] Of relevance to the latter proposition, GPI-linked CD14 may localise in detergent-insoluble lipid microdomains rich in a variety of signalling molecules [76]. Since GPI-linked and recombinant transmembrane forms of CD14 can both signal pro-inflammatory responses following engagement with LPS, it seems that such microdomain localisation of CD14 is not required for inflammatory signalling [77]. However, certain signals (increased intracellular Ca^{2+}, for example) consequent to cross-linking of CD14 by antibodies appear specific to the GPI-linked, as distinct from the transmembrane, form of the molecule [77], suggesting that microdomain localisation plays a key role in signal transduction following CD14 ligation. The physiological counterparts of such cross-linking ligands are not known but it is conceivable that the complex, multivalent apoptotic cell presents an effective cross-linking ligand to GPI-linked CD14 generating distinct, microdomain-localised signals from those generated by LPS. Since effective microbial polymeric CD14 ligands elicit pro-inflammatory responses following binding to GPI-linked CD14 [62], high-level CD14 cross-linking would be expected to be insufficient (though it might be necessary) to promote engagement of non- or anti-inflammatory signalling pathways in the responding macrophage.

Given the complexity of the apoptotic cell as a potential ligand for CD14 and the indications, discussed above, that the preferential ligands for CD14 on apoptotic cells are probably glycoproteins, it seems most likely at the time of writing that, following interaction with apoptotic-cell-associated ligands, CD14 interacts in cis with other macrophage membrane proteins to initiate signalling rather than transducing signals itself. The β2 integrins CR3 (CD11b/CD18; Mac-1) and CR4 (CD11c/CD18; p150,95) appear to associate with GPI-linked receptors, including CD14. LPS/LBP binding to CD14 may direct physical interactions between CR3 and CD14 [73] that lead to downstream signalling mediated via the cytoplasmic tails of the associated integrin molecules. Recently, the transmembrane Toll-like receptors, notably TLR2 and TLR4, have also been implicated in signal transduction following binding of LPS to CD14 [78–80; paper by Kitchens, this vol.]. The existence of a surface component of T cells that can transduce negative regulatory signals following binding to soluble CD14 lends further support for CD14-associating signalling complexes [61]. Recent work also indicates that GPI-linked CD14 associates physically with src family kinases and Gi/Go heterotrimeric G proteins [81]. The latter appear capable of regulating CD14-mediated pro-inflammatory cytokine production.

The CD14-mediated response of macrophages to apoptotic cells is non-inflammatory [1, 22, 82] or anti-inflammatory (see below). This poses the question, therefore, of how GPI-linked CD14 can direct the diammetrically opposed pro-inflammatory and non-anti-inflammatory responses of macrophages to ligands such as LPS, and to apoptotic cells, respectively. One possibility is that the basis for the differential CD14-dependent responses of macrophages to LPS and to apoptotic cells lies in the identity of the transmembrane 'partner' protein(s) that associates with CD14 in each case, with the CD14-ligand determining the CD14-associating 'partner' molecules and consequently the signal-transduction complex. In this respect it is worthwhile noting the differential involvement of the complement receptors CR3 and CR4 in pro-inflammatory responses to LPS as compared with group B streptococcal cell walls [83]. An alternative view is that CD14-dependent pro-inflammatory macrophage responses to LPS are primarily governed by lipid transfer and turnover [75] and that CD14-dependent non- or anti-inflammatory responses to apoptotic cells are mediated by a profoundly different means, which could involve transmembrane CD14-signalling partners.

It has long been accepted that the phagocytic clearance of apoptotic cells is characteristically non-phlogistic, indeed apoptosis stimuli (such as X and UV irradiation) themselves have anti-inflammatory properties. Recent work has begun to unravel how apoptotic cells can elicit such non-phlogistic responses in the phagocytes that recognise them and a growing body of evidence indicates that the apoptotic cell actively suppresses inflammatory responses of the macrophage. The in vitro release of pro-inflammatory cytokines TNF-α, IL-12 and IL-1β by human monocytes stimulated with LPS has been observed to be significantly inhibited by apoptotic lymphocytes in the cultures [84]. By contrast, in the same experimental system, release of the anti-inflammatory cytokine IL-10 is stimulated by apoptotic cells. Similar observations on inhibition by apoptotic cells of the spontaneous or LPS- and zymosan-induced release of TNF-α and IL-1β from human macrophages have been made during independent studies of human macrophages phagocytosing apoptotic neutrophils or lymphocytes [23]. In the latter experiments, IL-8, GM-CSF and IL-10 release were also inhibited by apoptotic cells, whereas release of the anti-inflammatory cytokine TGF-β1 was enhanced. In the same study, phagocytosis of apoptotic cells was observed to inhibit release of thromboxane B_2 and LTC_4 and stimulate production of PGE_2 and PAF [23]. Furthermore, PAF and/or PGE_2 may stimulate TGF-β release, suggesting that TGF-β may play a pivotal anti-inflammatory role during apoptotic-cell clearance.

While further studies are required to extend these initial observations, a model is emerging in which apoptotic-cell clearance elicits a soluble factor-mediated feedback loop through which monocyte/macrophage pro-inflammatory cytokine release is actively inhibited. The involvement of particular macrophage receptors for apoptotic cells in this model awaits further experimentation, although some of the initial work suggested that CD36 could mediate anti-inflammatory effects following apoptotic-cell ligation since the inhibition of release of TNF-α, IL-12 and IL-1β together with the stimulation of IL-10 release by apoptotic cells were blocked by a mAb recognising the TSP-binding site of CD36 [84]. How CD14 in its role as a macrophage receptor for apoptotic cells fits into this picture is not yet clear since no direct experimentation has yet been undertaken. It would appear, however, that human monocyte-derived macrophages, which are known to clear apoptotic cells via a CD14 component [1, 54], are signalled by apoptotic cells to activate anti-inflammatory pathways that are dominant over pro-inflammatory signals, including LPS [23]. It might be predicted, therefore that anti-inflammatory CD14-generated signals are dominant over pro-inflammatory CD14-generated signals in macrophages. Testing this prediction will be an important avenue of future investigation.

Conclusions

Non-phlogistic clearance of apoptotic cells via macrophage surface CD14 requires a profound divergence in intracellular signalling from that elicited by CD14's pro-inflammatory ligands such as LPS. The capacity for the GPI-linked form of CD14 to signal differentially following ligand-binding is by no means without precedent. For example, stimulation of human monocytes with LPS elicits, not only early pro-inflammatory cytokine release, but also late, PTK-dependent release of the anti-inflammatory cytokine, IL-10 [85]. Others have shown that both pro- and anti-inflammatory cytokine production can be induced by *Borrelia burgdorferi* lipoproteins following binding to monocyte CD14 [86]. In addition, the differentiation stage of the myeloid cell can determine the outcome of GPI-linked CD14 signalling. Thus monocyte-derived macrophages undergo chemotaxis following interaction of lipoarabinomannan (LAM) with CD14. However an alternative response to LAM involving a rise in intracellular Ca^{2+} is observed following synchronous engagement of CD14 and the macrophage mannose receptor [87]. The requirement for differentiation stage-specific surface receptors and/or intracellular signalling molecules is a requirement of CD14-mediated apoptotic-cell clearance, since monocytes apparently fail to phagocytose apoptotic cells and bodies. Furthermore, COS cells expressing substantially higher levels of CD14 than macrophages are markedly less effective than macrophages in apoptotic-cell clearance [1]. This observation accords with the notion that COS cells lack the appropriate or optimal CD14-'partner' molecules required for efficient signal transduction following apoptotic-cell-associated ligand binding.

Continued progress in understanding the molecular mechanisms underlying the activation of macrophages following pro-inflammatory ligand binding to GPI-linked CD14 is likely to provide important clues to dissecting the anti-inflammatory signalling pathways that mediate apoptotic-cell clearance. Many additional questions remain: To what extent does CD14 direct non-phlogistic responses to apoptotic cells in vivo? What is the relationship between CD14-dependent apoptotic-cell clearance and other established modes of removal of apoptotic cells? How conserved is the process? How does the anti-inflammatory signalling that accompanies apoptotic-cell clearance by macrophages compare to that which is stimulated by intracellular parasites of macrophages? Future progress in this area will yield information not only of importance to our understanding of inflammation and apoptosis, but also of relevance to the improved treatment of inflammatory diseases and of diseases in which apoptosis is dysfunctional.

References

1   Devitt A, Moffatt OD, Raykundalia C, Capra JD, Simmons DL, Gregory CD: Human CD14 mediates recognition and phagocytosis of apoptotic cells. Nature 1998;392:505–509.
2   Thompson CB: Apoptosis in the pathogenesis and treatment of disease. Science 1995;267:1456–1462.
3   Kerr JFR, Wyllie AH, Currie AR: Apoptosis: A basic biological phenomenon with wide-ranging implications in tissue kinetics. Br J Cancer 1972;26:239–257.
4   Wyllie AH, Kerr JFR, Currie AR: Cell death: The significance of apoptosis. Int Rev Cytol 1980; 68:251–305.
5   Wyllie AH: Glucocorticoid-induced thymocyte apoptosis is associated with endogenous endonuclease activation. Nature 1980;284:555–556.
6   Oberhammer F, Wilson JW, Dive C, Morris ID, Hickman JA, Wakeling AE, Walker PR, Sikorska M: Apoptotic death in epithelial cells: Cleavage of DNA to 300 and/or 50 kb fragments prior to or in the absence of internucleosomal fragmentation. EMBO J 1993;9:3679–3684.
7   Enari M, Sakahira H, Yokoyama H, Okawa K, Iwamatsu A, Nagata S: A caspase-activated DNase that degrades DNA during apoptosis, and its inhibitor ICAD. Nature 1998;391:43–50.
8   Liu XS, Zou H, Slaughter C, Wang XD: DFF, a heterodimeric protein that functions downstream of caspase-3 to trigger DNA fragmentation during apoptosis. Cell 1997;89:175–184.
9   Cohen GM: Caspases: The executioners of apoptosis. Biochem J 1997;326:1–16.
10 Stroh C, SchulzeOsthoff K: Death by a thousand cuts: An ever increasing list of caspase substrates. Cell Death Different 1998;5:997–1000.
11 Krammer PH: CD95 (APO-1/Fas)-mediated apoptosis: Live and let die. Adv Immunol 1999;71: 163–210.
12 Green DR, Reed JC: Mitochondria and apoptosis. Science 1998;281:1309–1312.
13 Liu XS, Li P, Widlak P, Zou H, Luo X, Garrard WT, Wang XD: The 40-kDa subunit of DNA fragmentation factor induces DNA fragmentation and chromatin condensation during apoptosis. Proc Natl Acad Sci USA 1998;95:8461–8466.
14 Sakahira H, Enari M, Nagata S: Cleavage of CAD inhibitor in CAD activation and DNA degradation during apoptosis. Nature 1998;391:96–99.
15 Adams JM, Cory S: The Bcl-2 protein family: Arbiters of cell survival. Science 1998;281:1322–1326.

16 Hall SE, Savill JS, Henson PM, Haslett C: Apoptotic neutrophils are phagocytosed by fibroblasts with participation of the fibroblast vitronectin receptor and involvement of a mannose/fucose-specific lectin. J Immunol 1994;153:3218–3227.

17 Finnemann SC, Bonilha VL, Marmorstein AD, RodriguezBoulan E: Phagocytosis of rod outer segments by retinal pigment epithelial cells requires alpha v beta 5 integrin for binding but not for internalization. Proc Natl Acad Sci USA 1997;94:12932–12937.

18 Shiratsuchi A, Umeda M, Ohba Y, Nakanishi Y: Recognition of phosphatidylserine on the surface of apoptotic spermatogenic cells and subsequent phagocytosis by sertoli cells of the rat. J Biol Chem 1997;272:2354–2358.

19 Bennett MR, Gibson DF, Schwartz SM, Tait JF: Binding and phagocytosis of apoptotic vascular smooth muscle cells is mediated in part by exposure of phosphatidylserine. Circ Res 1995;77: 1136–1142.

20 Hughes J, Liu YQ, VanDamme J, Savill J: Human glomerular mesangial cell phagocytosis of apoptotic neutrophils: Mediation by a novel CD36-independent vitronectin receptor thrombospondin recognition mechanism that is uncoupled from chemokine secretion. J Immunol 1997;158:4389–4397.

21 Savill J, Fadok V, Henson P, Haslett C: Phagocyte recognition of cells undergoing apoptosis. Immunol Today 1993;14:131–136.

22 Meagher LC, Savill JS, Baker A, Fuller RW, Haslett C: Phagocytosis of apoptotic neutrophils does not induce macrophage release of thromboxane-b2. J Leuk Biol 1992;52:269–273.

23 Fadok VA, Bratton DL, Konowal A, Freed PW, Westcott JY, Henson PM: Macrophages that have ingested apoptotic cells in vitro inhibit proinflammatory cytokine production through autocrine/ paracrine mechanisms involving TGF-β, PGE2, and PAF. J Clin Invest 1998;101:890–898.

24 Ren Y, Savill J: Apoptosis: The importance of being eaten. Cell Death Different 1998;5:563–568.

25 Gregory CD: Phagocytic clearance of apoptotic cells: Food for thought. Cell Death Different 1998; 5:549–550.

26 Botto M, DellAgnola C, Bygrave AE, Thompson EM, Cook HT, Petry F, Loos M, Pandolfi PP, Walport MJ: Homozygous C1q deficiency causes glomerulonephritis associated with multiple apoptotic bodies. Nat Genet 1998;19:56–59.

27 Herrmann M, Voll RE, Zoller OM, Hagenhofer M, Ponner BB, Kalden JR: Impaired phagocytosis of apoptotic cell material by monocyte-derived macrophages from patients with systemic lupus erythematosus. Arthritis Rheum 1998;41:1241–1250.

28 Morris RG, Hargreaves AD, Duvall E, Wyllie AH: Hormone-induced cell death. 2. Surface changes in thymocytes undergoing apoptosis. Am J Pathol 1984;115:426–436.

29 Duvall E, Wyllie AH, Morris RG: Macrophage recognition of cells undergoing programmed cell death (apoptosis). Immunology 1985;56:351–358.

30 Dini L, Autuori F, Lentini A, Oliverio S, Piacentini M: The clearance of apoptotic cells in the liver is mediated by the asialoglycoprotein receptor. FEBS Lett 1992;296:174–178.

31 Falasca L, Bergamini A, Serafino A, Balabaud C, Dini L: Human kupffer cell recognition and phagocytosis of apoptotic peripheral blood lymphocytes. Exp Cell Res 1996;224:152–162.

32 Dini L, Carla EC: Hepatic sinusoidal endothelium heterogeneity with respect to the recognition of apoptotic cells. Exp Cell Res 1998;240:388–393.

33 Koopman G, Reutelingsperger CPM, Kuijten GAM, Keehnen RMJ, Pals ST, Vanoers MHJ: Annexin-V for flow cytometric detection of phosphatidylserine expression on B cells undergoing apoptosis. Blood 1994;84:1415–1420.

34 Verhoven B, Schlegel RA, Williamson P: Mechanisms of phosphatidylserine exposure, a phagocyte recognition signal, on apoptotic T lymphocytes. J Exp Med 1995;182:1597–1601.

35 Bratton DL, Fadok VA, Richter DA, Kailey JM, Guthrie LA, Henson PM: Appearance of phosphatidylserine on apoptotic cells requires calcium-mediated nonspecific flip-flop and is enhanced by loss of the aminophospholipid translocase. J Biol Chem 1997;272:26159–26165.

36 Fadok VA, Bratton DL, Frasch SC, WArner ML, Henson PM: The role of phosphatidylserine in recognition of apoptotic cells by phagocytes. Cell Death Different 1998;5:551–562.

37 Fadok VA, Voelker DR, Campbell PA, Cohen JJ, Bratton DL, Henson PM: Exposure of phosphatidylserine on the surface of apoptotic lymphocytes triggers specific recognition and removal by macrophages. J Immunol 1992;148:2207–2216.

38   Fadok VA, Savill JS, Haslett C, Bratton DL, Doherty DE, Campbell PA, Henson PM: Different populations of macrophages use either the vitronectin receptor or the phosphatidylserine receptor to recognize and remove apoptotic cells. J Immunol 1992;149:4029–4035.

39   Martin SJ, Reutelingsperger CPM, McGahon AJ, Rader JA, van Schie RCAA, LaFace DM, Green DR: Early redistribution of plasma membrane phosphatidylserine is a general feature of apoptosis regardless of the initiating stimulus: Inhibition by overexpression of Bcl-2 and Abl. J Exp Med 1995;82:1545–1556.

40   Martin SJ, Finucane DM, AmaranteMendes GP, Obrien GA, Green DR: Phosphatidylserine externalization during CD95-induced apoptosis of cells and cytoplasts requires ICE/CED-3 protease activity. J Biol Chem 1996;271:28753–28756.

41   Martin SJ, Obrien GA, Nishioka WK, McGahon AJ, Mahboubi A, Saido TC, Green DR: Proteolysis of fodrin (nonerythroid spectrin) during apoptosis. J Biol Chem 1995;270:6425–6428.

42   Zhuang JG, Ren Y, Snowden RT, Zhu HJ, Gogvadze V, Savill JS, Cohen GM: Dissociation of phagocyte recognition of cells undergoing apoptosis from other features of the apoptotic program. J Biol Chem 1998;273:15628–15632.

43   Savill J, Hogg N, Ren Y, Haslett C: Thrombospondin cooperates with CD36 and the vitronectin receptor in macrophage recognition of neutrophils undergoing apoptosis. J Clin Invest 1992;90: 1513–1522.

44   Mevorach D, Mascarenhas JO, Gershov D, Elkon KB: Complement-dependent clearance of apoptotic cells by human macrophages. J Exp Med 1998;188:2313–2320.

45   Moffatt OD, Devitt A, Bell ED, Simmons DL, Gregory CD: Macrophage recognition of ICAM-3 on apoptotic leukocytes. J Immunol 1999;162:6800–6810.

46   de Fougerolles AR, Springer TA: Intercellular adhesion molecule 3, a third adhesion counter-receptor for lymphocyte function-associated molecule 1 on resting lymphocytes. J Exp Med 1992; 175:185–190.

47   Campanero MR, Delpozo MA, Arroyo AG, Sanchez-Mateos P, Hernandez-Caselles T, Craig A, Pulido R, Sanchez-Madrid F: ICAM-3 interacts with LFA-1 and regulates the LFA-1/ICAM-1 cell adhesion pathway. J Cell Biol 1993;123:1007–1016.

48   Hernandez-Caselles T, Rubio G, Campanero MR, Delpozo MA, Muro M, Sanchez-Madrid F, Aparicio P: ICAM-3, the third LFA-1 counter-receptor, is a costimulatory molecule for both resting and activated T lymphocytes. Eur J Immunol 1993;23:2799–2806.

49   Griffiths CEM, Railan D, Gallatin WM, Cooper KD: The ICAM-3/LFA-1 interaction is critical for epidermal Langerhans cell alloantigen presentation to CD4(+) T cells. Br J Dermatol 1995; 133:823–829.

50   Starling GC, McLellan AD, Egner W, Sorg RV, Fawcett J, Simmons DL, Hart DNJ: Intercellular adhesion molecule 3 is the predominant costimulatory ligand for leukocyte function antigen 1 on human blood dendritic cells. Eur J Immunol 1995;25:2528–2532.

51   Savill J, Dransfield I, Hogg N, Haslett C: Vitronectin receptor-mediated phagocytosis of cells undergoing apoptosis. Nature 1990;343:170–173.

52   Luciani MF, Chimini G: The ATP binding cassette transporter ABC1, is required for the engulfment of corpses generated by apoptotic cell death. EMBO J 1996;15:226–235.

53   Platt N, Suzuki H, Kurihara Y, Kodama T, Gordon S: Role for the class A macrophage scavenger receptor in the phagocytosis of apoptotic thymocytes in vitro. Proc Natl Acad Sci USA 1996;93: 12456–12460.

54   Flora PK, Gregory CD: Recognition of apoptotic cells by human macrophages: Inhibition by a monocyte/macrophage-specific monoclonal antibody. Eur J Immunol 1994;24:2625–2632.

55   Fadok VA, Warner ML, Bratton DL, Henson PM: CD36 is required for phagocytosis of apoptotic cells by human macrophages that use either a phosphatidylserine receptor or the vitronectin receptor (alpha(v)beta(3)). J Immunol 1998;161:6250–6257.

56   Albert ML, Sauter B, Bhardwaj N: Dendritic cells acquire antigen from apoptotic cells and induce class I restricted CTLs. Nature 1998;392:86–89.

57   Albert ML, Pearce SFA, Francisco LM, Sauter B, Roy P, Silverstein RL, Bhardwaj N: Immature dendritic cells phagocytose apoptotic cells via alpha(v)beta(5) and CD36, and cross-present antigens to cytotoxic T lymphocytes. J Exp Med 1998;188:1359–1368.

58 Ellis RE, Jacobson DM, Horvitz HR: Genes required for the engulfment of cell corpses during programmed cell death in *Caenorhabditis elegans.* Genetics 1991;129:79–94.

59 Vita N, Lefort S, Sozzani P, Reeb R, Richards R, Borysiewicz LK, Ferrara P, Labeta MO: Detection and biochemical characteristics of the receptor for complexes of soluble CD14 and bacterial lipopolysaccharide. J Immunol 1997;158:3457–3462.

60 Vasselon T, Pironkova R, Detmers PA: Sensitive responses of leukocytes to lipopolysaccharide require a protein distinct from CD14 at the cell surface. J Immunol 1997;159:4498–4505.

61 Nores JER, Bensussan A, Vita N, Stelter F, Arias MA, Jones M, Lefort S, Borysiewicz LK, Ferrara P, Labeta MO: Soluble CD14 acts as a negative regulator of human T cell activation and function. Eur J Immunol 1999;29:265–276.

62 Gregory CD, Devitt A: CD14 and apoptosis. Apoptosis 1999;4:11–20.

63 Yu B, Hailman E, Wright SD: Lipopolysaccharide binding protein and soluble CD14 catalyze exchange of phospholipids. J Clin Invest 1997;99:315–324.

64 Wang PY, Kitchens RL, Munford RS: Phosphatidylinositides bind to plasma membrane CD14 and can prevent monocyte activation by bacterial lipopolysaccharide. J Biol Chem 1998;273:24309–24313.

65 Soell M, Lett E, Holveck F, Scholler M, Wachsmann D, Klein JP: Activation of human monocytes by streptococcal rhamnose glucose polymers is mediated by CD14 antigen, and mannan-binding protein inhibits TNF-alpha release. J Immunol 1995;154:851–860.

66 Cavaillon JM, Marie C, Caroff M, Ledur A, Godard I, Poulain D, Fitting C, Haeffner-Cavaillon N: CD14/LPS receptor exhibits lectin-like properties. J Endotoxin Res 1996;3:471–480.

67 Weidemann B, Schletter J, Dziarski R, Kusumoto S, Stelter F, Rietschel ET, Flad HD, Ulmer AJ: Specific binding of soluble peptidoglycan and muramyldipeptide to CD14 on human monocytes. Infect Immun 1997;65:858–864.

68 Balasubramanian K, Schroit AJ: Characterization of phosphatidylserine-dependent beta(2)-glycoprotein I macrophage interactions: Implications for apoptotic cell clearance by phagocytes. J Biol Chem 1998;273:29272–29277.

69 de Fougerolles AR, Diamond MS, Springer TA: Heterogenous glycosylation of ICAM-3 and lack of interaction with mac-1 and p150,95. Eur J Immunol 1995;25:1008–1012.

70 Pugin J, Heumann D, Tomasz A, Kravchenko VV, Akamatsu Y, Nishijima M, Glauser MP, Tobias PS, Ulevitch RJ: CD14 is a pattern recognition receptor. Immunity 1994;1:509–516.

71 Savill J: Apoptosis: Phagocytic docking without shocking. Nature 1998;392:442–443.

72 Ulevitch RJ, Tobias PS: Receptor-dependent mechanisms of cell stimulation by bacterial endotoxin. Ann Rev Immunol 1995;13:437–457.

73 Petty HR, Todd RF: Integrins as promiscuous signal transduction devices. Immunol Today 1996;17:209–212.

74 Joseph CK, Wright SD, Bornmann WG, Randolph JT, Kumar ER, Bittman R, Liu J, Kolesnick RN: Bacterial lipopolysaccharide has structural similarity to ceramide and stimulates ceramide-activated protein kinase in myeloid cells. J Biol Chem 1994;269:17606–17610.

75 Thieblemont N, Thieringer R, Wright SD: Innate immune recognition of bacterial lipopolysaccharide: Dependence on interactions with membrane lipids and endocytic movement. Immunity 1998;8:771–777.

76 Cebecauer M, Cerny J, Horejsi V: Incorporation of leucocyte GPI-anchored proteins and protein tyrosine kinases into lipid-rich membrane domains of COS-7 cells. Biochem Biophys Res Commun 1998;243:706–710.

77 Pugin J, Kravchenko VV, Lee JD, Kline L, Ulevitch RJ, Tobias PS: Cell activation mediated by glycosylphosphatidylinositol-anchored or transmembrane forms of CD14. Infect Immun 1998;66:1174–1180.

78 Yang RB, Mark MR, Gray A, Huang A, Xie MH, Zhang M, Goddard A, Wood WI, Gurney AL, Godowski PJ: Toll-like receptor-2 mediates lipopolysaccharide-induced cellular signalling. Nature 1998;395:284–288.

79 Poltorak A, He XL, Smirnova I, Liu MY, VanHuffel C, Du X, Birdwell D, Alejos E, Silva M, Galanos C, Freudenberg M, RicciardiCastagnoli P, Layton B, Beutler B: Defective LPS signaling in C3H/HeJ and C57BL/10ScCr mice: Mutations in Tlr4 gene. Science 1998;282:2085–2088.

80 Wright SD: Toll, a new piece in the puzzle of innate immunity. J Exp Med 1999;189:605–609.
81 Solomon KR, KurtJones EA, Saladino RA, Stack AM, Dunn IF, Ferretti M, Golenbock D, Fleisher GR, Finberg RW: Heterotrimeric G proteins physically associated with the lipopolysaccharide receptor CD14 modulate both in vivo and in vitro responses to lipopolysaccharide. J Clin Invest 1998;102:2019–2027.
82 Stern M, Savill J, Haslett C: Human monocyte-derived macrophage phagocytosis of senescent eosinophils undergoing apoptosis: Mediation by alpha(v)beta(3)/cd36/thrombospondin recognition mechanism and lack of phlogistic response. Am J Pathol 1996;149:911–921.
83 Medvedev AE, Flo T, Ingalls RR, Golenbock DT, Teti G, Vogel SN, Espevik T: Involvement of CD14 and complement receptors CR3 and CR4 in nuclear factor-kappa B activation and TNF production induced by lipopolysaccharide and group B streptococcal cell walls. J Immunol 1998; 160:4535–4542.
84 Voll RE, Herrmann M, Roth EA, Stach C, Kalden JR, Girkontaite I: Immunosuppressive effects of apoptotic cells. Nature 1997;390:350–351.
85 Meisel C, Vogt K, Platzer C, Randow D, Liebenthal C, Volk HD: Differential regulation of monocytic tumor necrosis factor-alpha and interleukin-10 expression. Eur J Immunol 1996;26:1580–1586.
86 Giambartolomei GH, Dennis VA, Lasater BL, Philipp MT: Induction of pro- and anti-inflammatory cytokines by *Borrelia burgdorferi* lipoproteins in monocytes is mediated by CD14. Infect Immun 1999;67:140–147.
87 Bernardo J, Billingslea AM, Blumenthal RL, Seetoo KF, Simons ER, Fenton MJ: Differential responses of human mononuclear phagocytes to mycobacterial lipoarabinomannans: Role of CD14 and the mannose receptor. Infect Immun 1998;66:28–35.

Christopher D. Gregory, Institute of Cell Signalling and School of Biomedical Sciences, D Floor, University of Nottingham Medical School, Queen's Medical Centre,
Nottingham, NG7 2UH (UK)
Tel. +44 115 970 9370, Fax +44 115 970 9926, E-Mail Chris.Gregory@nottingham.ac.uk

Jack RS (ed): CD14 in the Inflammatory Response.
Chem Immunol. Basel, Karger, 2000, vol 74, pp 141–161

····················

TNF in the Inflammatory Response

Daniela N. Männel, Bernd Echtenacher

Institut für Pathologie, Tumorimmunologie, Universität Regensburg, Deutschland

Initial observations that cancer patients sometimes respond to severe infection with regression of their tumors led to the intentional administration of mixtures of bacterial extracts to patients with inoperable neoplastic diseases [1]. Many years of experimental and clinical research resulted in the isolation of the therapeutically active fraction from the bacterial mixtures which turned out to be endotoxin, the bacterial lipopolysaccharide (LPS), derived from the cell wall of gram-negative bacteria. Based on these findings, LPS was considered to be a model substance which mimics biological activities observed in sepsis [2]. In addition, serum of mice sensitized to LPS by infection with bacillus Calmette-Guérin (BCG) and then challenged with LPS was found to contain a host mediator which produces hemorrhagic and coagulative necrosis in tumors of recipient mice [3,4]. This host mediator was termed 'tumor necrosis factor' (TNF) and hopes for an effective tumor therapy with TNF spurred further work.

In 1985, the TNF gene was cloned and TNF expressed in *Escherichia coli* [5]. Independently, a mediator conferring severe wasting and cachexia observed in chronic infection, named 'cachectin', was discovered which turned out to be identical with TNF [6, 7]. Detailed studies of the biological properties of TNF led to the conclusion that TNF is the principal endogenous mediator of endotoxic shock, having pleiotropic biological activities of so great a degree as to explain its numerous physiological and pathological functions in infection and inflammation [8, 9]. Today, the accumulated evidence suggests that the tumor-necrotizing activity of TNF, in fact, is an epiphenomenon which at present is only left to be exploited in locoregional treatment of cancer patients such as isolated limb perfusions [10]. Additionally, data in models for experimental metastasis showed that in contrast to the well-recognized antitumoral activity, TNF also promotes tumor metastasis which does not support the idea of TNF as an ideal anticancer treatment [11].

The recognition of TNF as a major mediator of the systemic manifestations in some models of septic shock initiated a number of 'anticytokine therapy' trials with the aim to prevent multiple organ damage by neutralizing TNF or blocking its action. However, also in this field, the clinical responses were disappointing [12]. The conflicting data from experimental research and clinical trials are probably due to the fact that the anticytokine therapy often starts too late, the biological changes in an ongoing sepsis are more complex than after a single bolus injection of LPS, and no definite parameter exists as yet to determine the actual immune status of a patient. Though TNF can be a powerful mediator promoting tissue damage it nevertheless remains an essential initiator and regulator of host defense. Because of this it remains an intriguing task to try to define therapeutic concepts which exploit its beneficial properties. In this overview we would like to concentrate on the complex and ambivalent roles of TNF in bacterial infections.

TNF Structure and Function

TNF biosynthesis and its mechanism of action are very well characterized. Human TNF is encoded by a single gene which was mapped to the middle of the major histocompatibility complex, on the short arm of chromosome 6 [13]. Production of TNF is strongly regulated. Among the regulatory elements responsible for transcriptional activation of TNF, NFκB-responsive elements play a central role in conferring LPS inducibility [14]. TNF is first expressed as a type II transmembrane protein (memTNF) of 26 kD which is proteolytically cleaved off the membrane by a metalloproteinase [15, 16] to form a soluble noncovalently linked homotrimer (solTNF). A mature TNF molecule consists of a nonglycosylated protein chain of 157 amino acids with a single internal disulfide bond. Both the solTNF homotrimer and the memTNF are biologically active [17]. The three-dimensional structure of the mature solTNF has been found to be a triangular pyramid with receptor binding sites located in the groove between the subunits on the lower half of the pyramid [18–21].

Injection of recombinant TNF leads to symptoms which are indistinguishable from LPS toxicity, supporting the notion that TNF constitutes one of the most prominent mediators induced by LPS [22]. The important biological function of TNF as an inflammatory cytokine has been further emphasized with TNF transgenic mice which develop chronic inflammatory reactions like arthritis [23]. TNF is one of the major secretory products of LPS-activated macrophages but other cell types like granulocytes, T and B cells, NK cells and mast cells have also been identified as producers of TNF [24, 25]. Besides LPS, which represents the main stimulus, many other agents, including TNF itself act to induce TNF production. Disruption of the TNF gene by genetic deletion made mice highly susceptible to infectious agents and resistant to TNF toxicity and to lethality induced by minute amounts of LPS following D-galactosamine treatment [26, 27]. In addition, these mice had a deficiency in granuloma development and problems in limiting the extent and duration of the inflammatory process after *Propionibacterium acnes* injection, pointing to an important role of TNF in establishing homeostasis and repair following inflammatory injury.

Two receptors for TNF exist with molecular weights of 55 kD (p55TNFR) and 75 kD (p75TNFR), respectively. They are homologous in their extracellular domains but differ completely in their intracellular domains [28]. The conically shaped solTNF binds to the receptors with the tip pointing to the target cell, thus clustering with its three receptor binding sites the TNFR in the membrane [29]. After TNF binding or stimulation, e.g. with LPS, the p55TNFR is mostly internalized and partly, like the p75TNFR and TNF itself, proteolytically cleaved off to become a soluble TNF-binding protein in the plasma [30]. Activated granulocytes and monocytes seem to be the main producers of soluble TNF receptors, molecules which are capable of neutralizing TNF or of binding it for protracted release. High levels of soluble TNFR were found in patients with sepsis and the ratio of TNF to soluble p75TNFR correlated with MODS score and mortality [31, 32].

Interaction of TNF with its receptors activates a variety of cellular enzymes and transcription factors leading to activation, differentiation or death of the target cell [33, 34]. Among the pleiotropic effects of TNF its physiological role in inflammation has been more intensively investigated than its role in tissue remodelling and development [24]. The p55TNFR has a key function in mediating the inflammatory TNF effects [35]. This has been unequivocally demonstrated with p55TNFR-deficient mice. A multitude of deficiencies associated with TNF action such as disturbed organization of secondary lymphoid organs [36] and problems in clearing pathogenic bacteria became apparent in these mice. Mice deficient for the p55TNFR were unable to mount an early immune defense response to an infection with *Listeria monocytogenes* [37, 38] or *Mycobacterium tuberculosis* [39, 40]. In contrast, mice defective for the p75TNFR are relatively normal. They exhibit a slightly reduced sensitivity to the toxic effects of high doses of LPS, enhanced mortality after *L. monocytogenes* infection, and resistance to induction of TNF-induced skin necrosis [41], show diminished migration of Langerhans' cells and, therefore, reduced contact hypersensitivity reactions [42]. The only striking difference to normal controls is the lack of development of the cerebral malaria syndrome [43].

Only very few known TNF functions are exclusively mediated by the p75TNF, e.g. induction of thymocyte proliferation [44].

Both TNF receptors seem to cooperate: while the p55TNFR is sufficient for signalling, the p75TNFR amplifies the signal as shown for cytotoxicity [45], endothelial cell activation [46] and liver cell toxicity [47]. This has led to the 'ligand passing' hypothesis [48] which has been modified by more recent data showing that the p75TNFR has different binding characteristics for TNF [49]. Whereas the p55TNFR binds solTNF with high affinity, p75TNFR seems to be the better partner for memTNF [50]. Most cell types express both TNFR. The p55TNFR is rather stably expressed whereas the strongly regulated expression of the p75TNFR implies an important regulating function for this receptor in inflammation. Recent data with transgenic mice expressing the human p75TNFR demonstrate a hyperinflammatory syndrome even in the absence of both mouse TNF receptors or the ligand [51]. In addition, an intracellularly expressed TNF-binding molecule crossreacting with the p75TNFR has been described [52].

TNF as an Endotoxic Mediator

TNF in Experimental Sepsis

The first report of prevention of death from endotoxin shock in animals by TNF neutralization [53] was reproduced in several other animal models, including the protection of baboons after i.v. infusion of live bacteria [54–57], and was backed up by the finding that injection of recombinant TNF causes tissue injury and lethal shock [54, 58]. Intensive experimental studies revealed a critical role for TNF in the pathology associated with the systemic inflammatory response syndrome (SIRS). IL-1 which is also released upon LPS stimulation strongly potentiates the lethal effects of TNF [59]. D-Galactosamine dramatically sensitizes animals to the toxic effect of LPS or TNF acting via the p55TNFR [37, 60–62]. The mechanism of this pathological effect is acute liver failure due to transcriptional arrest of hepatocytes induced by D-galactosamine which leads to hepatocyte apoptosis upon exposure to TNF [63]. The fact that p55TNFR-deficient mice are completely protected from this TNF-induced liver toxicity whereas they are as sensitive to a lethal dose of LPS as wild-type mice makes it clear that LPS induces destructive processes that are independent of TNF-mediated toxicity [40].

TNF in Human Sepsis

In humans, circulating TNF was detected after LPS administration [64] and comparable metabolic responses were observed upon TNF and LPS infusions [65]. A remarkable correlation between plasma TNF levels and mortality of patients with meningococcal infections was found [66]. In several other clinical studies serum TNF concentrations and severity of infections were also found to be positively correlated [67–69]. Stimulation with LPS or TNF activates the DNA-binding factor NFκB which initiates transcription from immediate early genes in a wide variety of cells [70]. Nuclear extracts from peripheral blood mononuclear cells from nonsurviving sepsis patients showed an increased course of nuclear binding activity of NFκB compared with survivors [71], fitting the hypothesis of TNF as a central regulator of the inflammatory reactions in sepsis.

Anti-TNF Studies

With these findings in mind, it was surprising that treatment of sepsis or septic shock patients with TNF-neutralizing agents did not result in protection [12]. One study even showed a clear increase in mortality when sepsis patients were treated with a high dose of a soluble fusion protein consisting of the extracellular domain of the p75TNFR and the Fc portion of human IgG1 [72]. This discrepancy between experimental data and clinical findings may have two main causes: (1) The conclusion that TNF blockade would protect from septic shock was drawn mostly from the use of i.v. bolus administration of LPS and sometimes live bacteria as models for septic shock. (2) By a classical definition sepsis is the continuous or intermittent release of bacteria from a focus of infection into the bloodstream leading to symptoms of disease. For clinical purposes a more recent definition was proposed according to which sepsis is the systemic inflammatory response (SIRS) to infection, severe sepsis is sepsis associated with organ dysfunction, hypoperfusion or hypertension, and septic shock is a sepsis-induced refractory shock [73]. Very clearly, according to both definitions, it is wrong to call LPS shock a septic shock since no infection is involved. Furthermore, although LPS was described to be released from gram-negative bacteria during sepsis especially after antimicrobial treatment [74, 75], there is frequently no correlation between serum LPS levels and the severity of sepsis. In addition, the use of ex vivo whole blood cultures from patients infected with various gram-negative bacteria often failed to demonstrate a correlation between antimicrobial treatment, LPS release and TNF production [76–78]. In mice, after cecal ligation and puncture, LPS and TNF concentrations were not significantly different in antibiotic-treated mice compared to controls [79]. One of the rare exceptions where antimicrobial treatment clearly induces potentially fatal TNF production leading to sudden fever, rigors and persistent hypotension is the Jarisch-Herxheimer reaction in patients infected with *Borellia recurrentis* [80]. The predominant TNF-inducing bacterial factor in this case is the variable major protein of *B. recurrentis* [81].

Taking these considerations into account, it seems obvious that focal models of infection like cecal ligation and puncture (CLP) certainly represent more closely the clinical situation of sepsis than bolus injections of LPS or bacteria and, therefore, are better suited to study sepsis and the role of TNF as mediator of innate immunity [82].

TNF as an Antibacterial Host-Defense Mediator

From LPS to TNF

A probable sequence of events in sepsis is that bacteria or their products bind to LBP, this complex translocates to CD14, which signals via TLR4, e.g. into macrophages to produce TNF and other mediators to recruit host defense, e.g. neutrophils. In fact, op/op mice that due to a mutated CSF-1 gene have less than 5% of the normal macrophage numbers in the peritoneal cavity, produce less LPS-induced TNF and are very sensitive to E. coli-induced peritonitis. Death is preceded by a delayed and impaired recruitment of neutrophils to the site of infection [83]. Likewise, mice with a defective TLR4, e.g. C3H/HeJ or C57BL/10ScCr mice, are as resistant to LPS as they are susceptible to lethal infection with certain strains of gram-negative bacteria (Salmonella typhimurium, E. coli) [84–86]. Also LBP-deficient mice are more sensitive to S. typhimurium than control mice and LPS induces a much lower TNF serum response [87]. Surprisingly, the only available in vivo infection data about CD14-deficient mice do not fit into this scheme: mice lacking CD14 were, as expected, more resistant to LPS shock than control mice; however, after E. coli-induced peritonitis, not only serum TNF levels but also mortality and bacteremia were drastically reduced in these mice [88]. The reaction of these mice in other infection models should be revealing because of the important antibacterial role CD14 was predicted to play on the basis of in vitro and in vivo data [89]. To add to the confusion, despite the fact that expression of LBP mRNA and CD14 mRNA is enhanced in normal mice after CLP [90] both LBP-deficient mice [own unpubl. data] and TLR4-deficient mice [91; and own unpubl. data] show identical survival rates after CLP as the respective control mice. Thus, the observed enhanced susceptibility to gram-negative bacteria of these mice cannot be generalized to all kinds of sepsis. A possible explanation for these data may be, that in contrast to a 'monomicrobial' infection, during the polymicrobial infection ensueing CLP other mechanisms activate, independently of LBP or TLR4, the TNF-dependent host defence.

TNF Protects in Experimental Sepsis

Many results from infection models document the importance of TNF-TNF receptor interaction for the outcome of bacterial infections. TNF-deficient mice can be protected from CLP-induced mortality by injection of TNF [92] and CLP-induced mortality is higher in mice after TNF neutralization [93]. Similarily, TNF administered 1–24 h before infection protects from normally lethal *Listeria* challenge and listeriosis is exacerbated if TNF is neutralized during the first 3 days of infection [94]. Mice deficient for TNF and mice lacking the p55TNFR are resistant to normally lethal LPS injections and are highly susceptible to the peritonitis induced by CLP [95]. Mice deficient for TNF and its p55TNFR also exhibit enhanced mortality in *E. coli*-induced peritonitis. In both genetically modified mouse strains mortality was accompanied by increased bacteremia. Interestingly, in the infected p55TNFR-deficient mice the blood levels of biologically active TNF were much higher compared to the infected wild-type mice [96] which could be due to the lack of circulating TNF-binding soluble p55TNFR. Lack of both TNF receptors exacerbates infection with *Brucella abortus* and leads to a reduced IL-12 response to these bacteria [97].

One function of TNF in antimicrobial defense could consist of supporting appropriate antigen presentation. Epidermal Langerhans' cells represent the major antigen-presenting dendritic cells within the skin. These cells capture antigen and migrate into regional lymph nodes where they present the antigen to B and T cells, thereby initiating the primary immune response. TNF has been shown to be importantly involved in dendritic cell migration and contact hypersensitivity [98, 99] and the interaction of TNF with the p75TNFR has been suggested to play an important role in promoting dendritic cell migration [42].

TNF Producer Cells

TNF is secreted when its major producers, monocytes and macrophages, are exposed to LPS, to dead or live bacteria, to antibody-antigen complexes and to complement components. Besides LPS and gram-negative bacteria also gram-positive bacteria, viruses, protozoa, and fungi induce TNF production [55, 56, 94, 100–102]. Since the identification of T lymphocytes as producers of deleterious levels of TNF in the toxic shock syndrome activated by bacterial superantigens, these cells have to be considered, too, as a possible TNF source in bacterial infections. Although bacteria capable of producing superantigens can cause lethal infections in mice [103], the action of superantigens does not seem to be a relevant pathomechanism in mouse models of septic shock unless these mice are sensitized with *D*-galactosamine. In the mixed bacterial infection after CLP, T cells do not seem to be important as contributors of protective TNF because the mortality of SCID mice which are devoid of mature B and T cells is not different from normal mice and can be strongly enhanced by TNF neutralization [93].

Macrophages are expert bactericidal cells and are considered to be the major TNF producers [104]. Like macrophages, neutrophils are also known as TNF producers and are essential for elimination of bacteria. It is, therefore, not easy to assess the importance of the TNF contribution of macrophages and neutrophils during infection independently of the direct impact of these cells on bacteria. As mentioned above, op/op mice with strongly reduced macrophage numbers produce less LPS-induced TNF and are very sensitive to *E. coli*-induced peritonitis [83]. Depletion of neutrophils with cyclophosphamide, antineutrophil antibodies or treatment of mice with anti-GCSF antibodies before CLP enhanced mortality [95].

Mast Cells as TNF Producers

Our initial obervation of decreased survival of mice after CLP if treated with anti-TNF antibodies within a period up to 8 h after surgery, induced the question about the source of this early TNF. Because the mortality was the higher the earlier after surgery TNF was neutralized, a rapid TNF producer should be involved. Peritoneal and other mast cells have preformed TNF stores which can be rapidly released within minutes upon stimulation. New TNF can be synthesized and released over a prolonged time with a TNF mRNA maximum at 30–60 min after stimulation [105, 106]. Mast-cell-deficient mice, indeed, proved to be extremely sensitive to CLP-induced mortality [107]. These mice are also very susceptible to intraperitoneal or pulmonary infection with *Klebsiella pneumoniae* [108]. They also succumb to lower numbers of orally administered *S. typhimurium* than normal littermates [109]. Normal resistence to CLP-induced peritonitis and infection with *K. pneumoniae* was restored when such mice were reconstituted with cultured mast cells, and TNF production in peritoneal cavities after *K. pneumoniae* infection depended on the presence of mast cells [107, 108]. Since, like in normal mice, protection conferred by mast cell reconstitution was abrogated by TNF neutralization, and TNF administration to mast-cell-deficient mice protected them from CLP-induced mortality, it is very likely that TNF is the critical mast cell mediator in this model. This question will be rigorously tested by reconstitution of mast-cell-deficient mice with mast cells from TNF-deficient mice.

Mast cells occur at sites where they can frequently encounter bacterial invaders. Besides the peritoneal cavity where serosal mast cells, after intestinal injury, will interact mainly with intestinal bacteria, mast cells are also found in the skin, the mucosa and submucosa of the genitourinary, gastrointestinal, and respiratory tract, and adjacent to lymphatic and blood vessels. There is no syndrome of increased susceptibility to infection known in humans correlating with mast cell deficiency. The fact that all vertebrates possess mast cells strongly points to an important role for mast cells in host defence [110, 111].

Mast cells do not express CD14 and it is therefore unlikely that LPS effects on mast cells are mediated via this pathway. However, there are many ways for mast cells to recognize the pathogens and the numerous mast-cell-activating mediators which indicate the presence of pathogens perceived by the innate or the adaptive immune system. The best-known stimulus for mast cell activation are IgE antigen complexes. After stimulation via the FcɛR mast cells release within minutes preformed TNF and other mediators like histamine [106] which makes the reported IgE responses to bacteria (*Helicobacter*, *S. aureus*) a likely part of the host's immune response against these pathogens [112–114]. Production of antibodies of the IgG class is one of the best-known efficient immune responses against bacteria. It is less well known that mast cells bear FcγRIII allowing them to release TNF as quickly and to the same extent as after stimulation via the FcɛR [115]. Thus, via its Fc-receptors mast cells become integrated into the adaptive immune system as important effector cells.

The dangers of integrating so impulsive a cell into the adaptive immune system become evident in anaphylaxis. Systemic anaphylaxis can be induced both via FcγRIII and via FcɛR. Although an anaphylactic reaction can be induced even in mast-cell-deficient mice, the weaker response exhibited by these mice compared to normal mice indicates the important role of mast cells [116, 117]. Monocytes, macrophages, neutrophils, and NK cells also express FcγRIII and are, therefore, likely to be responsible for the mast-cell-independent part of anaphylaxis. Because we regard anaphylaxis as an exaggerated and misguided immune response, which in a moderate form provides antibody-dependent host defence and because the cells involved are all potential TNF producers the question arises whether TNF is also crucially involved in anaphylaxis.

Stimuli for Early TNF Release

Binding and phagocytosis of bacteria with fimbria containing FimH (*E. coli, K. pneumoniae, Enterobacter cloacae*) depends on CD48, a receptor found on macrophages and also on mast cells [112, 118]. This event triggers TNF release from cultured and peritoneal mast cells [108, 119]. Further, bacterial infection and LPS induce complement activation either via the classical Ig-dependent pathway, the alternative pathway, or the C1q-like mannose-binding lectin (MBL). CR3-expressing cells, including mast cells, can directly bind and phagocytose opsonized bacteria whereby they become activated [120–122]. Accordingly, neutrophil influx and TNF release by mast cells in the peritoneal reverse passive Arthus reaction is complement-dependent as shown in decomplemented mice [123]. Mice deficient for C3 or C4 are highly susceptible to endotoxin shock [124], streptococcal infection [125], and CLP-induced peritonitis [126]. In the CLP model this has been shown to be accompanied by reduced peritoneal TNF levels and impaired mast cell activation. Administration of C3 improved both these parameters and survival [126]. These data demonstrate the importance of complement in activating TNF producer cells in bacterial peritonitis. That intact C3 and C4, however, are not sufficient for TNF production and survival of bacterial infections becomes evident with C5aR-deficient mice which are unable to clear *Pseudomonas aeroginosa* instilled into the lung [127]. The importance of C5a in IgG-induced production of TNF by mast cells was demonstrated when a reverse passive Arthus reaction in the peritoneal cavity of C5aR-deficient mice was initiated, and decreased levels of TNF occurred together with reduced migration of neutrophils [128]. Reduction of these two parameters may also be responsible for the increased susceptibilty of C5aR-deficient mice to CLP-induced peritonitis [own unpubl. data]. The most potent complement activator of all Ig classes are natural IgM antibodies which can lyse and opsonize bacteria. Like C4- or C3-deficient mice, mice deficient of soluble IgM showed increased susceptibility to CLP-induced mortality associated with decreased TNF levels in serum and peritoneal cavity, decreased neutrophil recruitment, and increased load of peritoneal bacteria. This makes it likely that IgM stimulates TNF production through the complement pathway, because IgM is not known to bind directly to TNF-producing cells [129].

Substance P (SP), a neurotransmitter, stimulates TNF secretion from mouse macrophages and mast cells [130]. TNF production and the inflammatory response stimulated by *Clostridium difficile* is diminished in the SP receptor-deficient mice [131]. LPS-induced TNF production is also reduced in mice if their SP receptors are blocked or if SP synthesis is inhibited [132]. We tested the hypothesis that SP administration might protect mice from septic peritonitis. Preliminary results indicate that this, indeed, is the case and that protection requires TNF and mast-cells, suggesting a connection between nociceptive C fibers as producers of SP and mast cell-derived TNF that augments neutrophil influx and exudation of plasma components into areas of inflammation. SP may also be the key molecule to explain preliminary data we obtained after lipid A injection into normal and mast-cell-deficient mice, namely, that after 30 min the mast cell-deficient mice exhibited 4-fold less serum TNF than normal mice. Although mast cells in vitro produce IL-6 upon LPS exposition, no report about in vitro TNF production after LPS stimulation can be found [133]. So, because in vivo LPS-induced TNF production is SP-dependent, LPS-induced SP may act on mast cells as SP-reactive TNF producers. This sequence of events could explain the early LPS-induced TNF production which is required for protection to CLP-induced peritonitis until other cells, e.g. macrophages, start to release TNF.

Functions of Early TNF

Intraperitoneal administration of TNF induces influx of neutrophils into the peritoneal cavity [134]. During *K. pneumoniae* infection of lung or peritoneal cavity, only mice with normal numbers of mast cells and, therefore, normal TNF production in lung and peritoneal cavity, revealed normal neutrophil influx, bacterial clearance and mouse survival [108]. Also immune complex-induced neutrophil influx into the peritoneal cavity and TNF production [115] as well as IgE-dependent cutaneous reactions, neutrophil infiltration and edema formation, are mast-cell-dependent and can be partially inhibited by TNF neutralization [135]. Thus, in infections an essential function of the early mast-cell-derived TNF and TNF from other sources is the rapid recruitment and activation of neutrophils. In addition, TNF-dependent and -independent exudation brings many plasma components at the site of infection that bind to bacteria and/or bacterial products and further enhance TNF production, like complement, IgM, IgG, MBL [136], LBP, sCD14 or, as in the case of fibrinogen, provide the substrate for fibrin formation that starts localization of the infection.

TNF exerts a well-known procoagulant activity. In the CLP sepsis model neutralization of TNF inhibits formation of peritoneal adhesions, the first step towards localization of the infection, and enhances mortality [137]. Under these circumstances the procoagulant activity of TNF seems to be essential because all measures of prevention of adhesions or dissolution of adhesions with anticoagulants or the fibrinolytic agent urokinase, respectively, increase mortality, too [own unpubl. data]. The most potent inducer of peritoneal adhesions among the intestinal bacteria was found to be *Bacteroides fragilis*. Its capsular polysaccharides promote adhesions of these bacteria to mesothelial cells and peritoneal macrophages as well as the formation of peritoneal adhesions. These can be inhibited by TNF neutralization [138]. In addition to these experimental data, after surgical intraabdominal manipulations in patients higher grades of adhesions correlated with higher levels of TNF both in serum and peritoneal exudate [139].

The role of TNF and other cytokines in alteration of coagulation and fibrinolysis in human sepsis and various models of sepsis and endotoxemia is not yet clear. A common view of causes and effects in sepsis is as follows: During sepsis bacteria and bacterial products (e.g. LPS) trigger production of cytokines (TNF, IL-1, IL-6, etc.) which contribute, via tissue factor induction, to activation of the coagulation cascade resulting in fibrin deposition, disseminated intravascular coagulation (DIC), and organ failure. In parallel, cytokine-enhanced expression of plasminogen activator inhibitor-1 (PAI-1) inhibits fibrinolysis, thereby increasing the rate of fibrin accumulation and DIC. There are many partially conflicting data about the role of TNF in these processes. While there is far-reaching agreement that TNF induces expression of tissue factor and PAI-1 and that TNF injection into humans and other species leads to virtually the same coagulation phenomena as after LPS injection [140], it became evident that neutralization of TNF after LPS injection in different species did not prevent activation of coagulation [141]. During endotoxemia in chimpanzees or humans, TNF-neutralization was reported not to influence activation of coagulation but to inhibit activation of fibrinolysis [142, 143]. However, in *E. coli*-induced bacteremia in chimpanzees coagulation activation was attenuated by TNF-neutralization while activation of fibrinolysis was unchanged and the release of PAI-1 was inhibited [144]. Even if one ignores the complexities of the relationship of TNF and coagulation, and tries to inhibit coagulation directly during sepsis, the results vary. Tissue factor pathway inhibitor (TFPI) treatment protected baboons from death after i.v. *E. coli* challenge [145, 146] but was unable to protect pigs from *E. coli*-laden fibrin clots implanted i.p. [147]. This, and the lethal consequences of coagulation inhibition after CLP sepsis, suggests that one very important function of the coagulation system is to localize infection. Because the cause of human sepsis usually is not i.v. infused bacteria, one should take into account this function when inhibition of coagulation is considered in sepsis.

TNF in Chronic Inflammatory Diseases

TNF plays an important yet undefined role in the pathogenesis of auto-immune diseases [148–151]. It is involved in chronic inflammatory diseases like rheumatoid arthritis and Crohn's disease. Neutralization of TNF improved Crohn's disease in a number of studies [152]. Severity of disease in patients with rheumatoid arthritis is positively correlated with TNF content of the synovial fluid and tissue, and TNF action blockade by treatment with an antibody to TNF or a p75TNFR-IgFc construct clearly improved the symptoms [153].

Interestingly, in experimental autoimmune encephalitis (EAE), a model for another autoimmune disease, multiple sclerosis, EAE did not resolve in TNF-deficient mice in contrast to normal mice. TNF deficiency even rendered EAE-resistant mice EAE susceptible. Moreover, TNF treatment of both normal and TNF-deficient mice reduced the severity of EAE [151]. The fact that in patients with multiple sclerosis increased disease activity was observed after TNF neutralization shows that neutralization of TNF also in human autoimmune disease is not simply beneficial [154]. The use of an experimental model which is closer to rheumatoid arthritis illustrates further the complicated role TNF plays in host protection and destruction: *S. aureus* infection both in humans and mice can cause polyarticular septic arthritis. At low infective doses of *S. aureus* more normal mice than mice deficient for TNF and lymphotoxin develop chronic arthritis. However, the gene-deficient mice die from doses that normal mice will survive [103]. The injection of killed *P. acnes* will serve as a last example for the ambivalent role of TNF in control of inflammation. In normal mice this leads to a rapid inflammation and subsequent resolution, in TNF-deficient mice, however, to a weak initial inflammatory response and after weeks to a vigorous, eventually lethal inflammatory response [27]. Thus, despite the promising results of TNF neutralization in the clinical studies with rheumatoid arthritis and Crohn's disease, the experimental data indicate that TNF is probably not a magic bullet for chronic inflammatory diseases.

Conclusion

It seems as if the role of TNF in the history of biomedical research is characterized by high expectations for therapy inspired by promising experimental findings, followed by disillusionment in the clinic. TNF therapy for tumor patients failed to live up to the high hopes placed on it. In a similar way, the great promise of anti-TNF strategies in sepsis was followed by disappointment at the results of the clinical trials. Equally, the current euphoria at the results of anti-TNF therapy in chronic inflammatory diseases may need to be revised in the light of longer clinical experience or new experimental findings. On the other hand, TNF research has led to a major revision of the way we think about mast cells. Mast cells, previously only blamed for mediating allergic reactions and anaphylaxis, are now recognised as playing an important role in host defence. Over the years, TNF has always been good for surprises and there is every reason to suppose that research on this molecule will remain as exciting in the future. The history of TNF research very nicely demonstrates how essential it is in science to keep an open mind and to constantly adjust one's view to new findings and, if necessary, to revise old and dearly loved dogmas.

References

1   Coley WB: The treatment of malignant tumor by repeated inoculations of erysipelas, with a report of 10 original cases. Am J Med Sci 1893;105:487.
2   Rietschel ET, Kirikae T, Schade FU, Mamat U, Schmidt G, Loppnow H, Ulmer AJ, Zähringer U, Seydel U, Di Padova F: Bacterial endotoxin: Molecular relationships of structure to activity and function. FASEB J 1994;8:217–225.

3   O'Malley WE, Achinstein B, Shear MJ: Journal of the National Cancer Institute, vol 29, 1962: Action of bacterial polysaccharide on tumors. II. Damage of sarcoma 37 by serum of mice treated with *Serratia marcescens* polysaccharide, and induced tolerance (classical article). Nutr Rev 1988; 46:389–391.

4   Carswell EA, Old LJ, Kassel RL, Green S, Fiore N, Williamson B: An endotoxin-induced serum factor that causes necrosis of tumor. Proc Natl Acad Sci USA 1975;72:3666.

5   Aggarwal BB, Kohr WJ, Hass PE, Moffat B, Spencer SA, Henzel WJ, Bringman TS, Nedwin GE, Goeddel DV, Harkin RN: TNF: Production, purifaction, characterization. J Biol Chem 1985;260: 2345–2354.

6   Beutler B, Greenwald D, Hulmes JD, Chang M, Pan YC, Mathison J, Ulevitch R, Cerami A: Identity of tumour necrosis factor and the macrophage-secreted factor cachectin. Nature 1985;316: 552–554.

7   Beutler B, Cerami A: The biology of cachectin/TNF-α primary mediator of the host response. Annu Rev Immunol 1989;7:625–655.

8   Beutler B, Cerami A: Cachectin and TNF as two sides of the same biological coin. Nature 1986; 320:584–588.

9   Beyaert R, Fiers W: Tumor necrosis factor and lymphotoxin; in Mire-Sluis AR, Thorpe R (eds): Cytokines. London, Academic Press, 1998, pp 335–360.

10   Lejeune F, Lienard D, Eggermont A, Schraffordt Koops H, Rosenkaimer F, Gerain J, Klaase J, Kroon B, Vanderveken J, Schmitz P: Administration of high-dose tumor necrosis factor alpha by isolation perfusion of the limbs: Rationale and results. J Infus Chemother 1995;5:73–81.

11   Hafner M, Orosz P, Krüger A, Männel DN: TNF promotes metastasis by impairing natural killer cell activity. Int J Cancer 1996;66:388–392.

12   Zeni F, Freeman B, Natanson C: Anti-inflammatory therapies to treat sepsis and septic shock: A reassessment. Crit Care Med 1997;25:1095–1100.

13   Spies T, Blanck G, Bresnahan M, Sands J, Strominger JL: A new cluster of genes within the human major histocompatibility complex. Science 1989;243:214–217.

14   Jongeneel CV: The TNF and lymphotoxin promotors; in Beutler B (ed): Tumor Necrosis Factors: The Molecules and Their Emerging Role in Medicine. New York, Raven Press, 1992, pp 539–559.

15   Black RA, Rauch CT, Kozlosky CJ, Peschon JJ, Slack JL, Wolfson MF, Castner BJ, Stocking KL, Reddy P, Srinivasan S, et al: A metalloproteinase disintegrin that releases tumour-necrosis factor-alpha from cells. Nature 1997;385:729–733.

16   Moss ML, Jin SL, Milla ME, Bickett DM, Burkhart W, Carter HL, Chen WJ, Clay WC, Didsbury JR, Hassler D, Hoffman CR, Kost TA, Lambert MH, Leesnitzer MA, McCauley P, McGeehan G, Mitchell J, Moyer M, Pahel G, Rocque W, Overton LK, Schoenen F, Seaton T, Su JL, Becherer JD: Cloning of a disintegrin metalloproteinase that processes precursor tumour-necrosis factor-alpha. Nature 1997;385:733–736.

17   Perez C, Albert I, DeFay K, Zachariades N, Gooding L, Kriegler M: A nonsecretable cell surface mutant of tumor necrosis factor (TNF) kills by cell-to-cell contact. Cell 1990;63:251–258.

18   Jones EY, Stuart DI, Walker NP: Structure of tumour necrosis factor. Nature 1989;338:225–228.

19   Eck MJ, Sprang SR: The structure of tumor necrosis factor-alpha at 2.6 Å resolution: Implications for receptor binding. J Biol Chem 1989;264:17595–17605.

20   Fiers W: Tumor necrosis factor: Characterization at the molecular, cellular and in vivo level. FEBS Lett 1991;285:199–212.

21   Banner DW, D'Arcy A, Janes W, Gentz R, Schoenfeld HJ, Broger C, Loetscher H, Lesslauer W: Crystal structure of the soluble human 55 kD TNF receptor-human TNF beta complex: Implications for TNF receptor activation. Cell 1993;73:431–445.

22   Beutler B (ed): Tumor Necrosis Factors: The Molecules and Their Emerging Role in Medicine. New York, Raven Press, 1992.

23   Keffer J, Probert L, Cazlaris H, Georgopoulos S, Kaslaris E, Kioussis D, Kollias G: Transgenic mice expressing human tumour necrosis factor: A predictive genetic model of arthritis. EMBO J 1991;10:4025–4031.

24   Vassalli P: The pathophysiology of tumor necrosis factors. Annu Rev Immunol 1992;10:411–452.

25 Sidhu RS, Bollon AP: Tumor necrosis factor activities and cancer therapy: A perspective. Pharmacol Ther 1993;57:79–128.

26 Eugster HP, Müller M, Karrer U, Car BD, Schnyder B, Eng VM, Woerly G, Le Hir M, Di Padova F, Aguet M, Zinkernagel R, Bluethmann H, Ryffel B: Multiple immune abnormalities in tumor necrosis factor and lymphotoxin-alpha double-deficient mice. Int Immunol 1996;8:23–36.

27 Marino MW, Dunn A, Grail D, Inglese M, Noguchi Y, Richards E, Jungbluth A, Wada H, Moore M, Williamson B, Basu S, Old LJ: Characterization of tumor necrosis factor-deficient mice. Proc Natl Acad Sci USA 1997;94:8093–8098.

28 Tartaglia LA, Goeddel DV: Two TNF receptors. Immunol Today 1992;13:151–153.

29 Bazzoni F, Beutler B: How do tumor necrosis factor receptors work? J Inflamm 1995;45:221–238.

30 van Tits LJ, Bemelmans MH, Steinshamn S, Waage A, Leeuwenberg JF, Buurman WA: Non-signaling functions of TNF-R75: Findings in man and mouse. Circ Shock 1994;44:40–44.

31 Calvano SE, van der Poll T, Coyle SM, Barie PS, Moldawer LL, Lowry SF: Monocyte tumor necrosis factor receptor levels as a predictor of risk in human sepsis. Arch Surg 1996;131:434–437.

32 Pellegrini JD, Puyana JC, Lapchak PH, Kodys K, Miller Graziano CL: A membrane TNF-alpha/TNFR ratio correlates to MODS score and mortality. Shock 1996;6:389–396.

33 Heller RA, Krönke M: Tumor necrosis factor receptor-mediated signaling pathways. J Cell Biol 1994;126:5–9.

34 Wallach D, Boldin M, Varfolomeev E, Beyaert R, Vandenabeele P, Fiers W: Cell death induction by receptors of the TNF family: Towards a molecular understanding. FEBS Lett 1997;410:96–106.

35 Thoma B, Grell M, Pfizenmaier K, Scheurich P: Identification of a 60-kD tumor necrosis factor (TNF) receptor as the major signal transducing component in TNF responses. J Exp Med 1990;172:1019–1023.

36 Pasparakis M, Alexopoulou L, Grell M, Pfizenmaier K, Bluethmann H, Kollias G: Peyer's patch organogenesis is intact yet formation of B lymphocyte follicles is defective in peripheral lymphoid organs of mice deficient for tumor necrosis factor and its 55-kDa receptor. Proc Natl Acad Sci USA 1997;94:6319–6323.

37 Rothe J, Lesslauer W, Lötscher H, Lang Y, Koebel P, Kontgen F, Althage A, Zinkernagel R, Steinmetz M, Bluethmann H: Mice lacking the tumour necrosis factor receptor 1 are resistant to TNF-mediated toxicity but highly susceptible to infection by Listeria monocytogenes. Nature 1993;364:798–802.

38 Pfeffer K, Matsuyama T, Kundig TM, Wakeham A, Kishihara K, Shahinian A, Wiegmann K, Ohashi PS, Krönke M, Mak TW: Mice deficient for the 55 kd tumor necrosis factor receptor are resistant to endotoxic shock, yet succumb to L. monocytogenes infection. Cell 1993;73:457–467.

39 Flynn JL, Goldstein MM, Chan J, Triebold KJ, Pfeffer K, Lowenstein CJ, Schreiber R, Mak TW, Bloom BR: Tumor necrosis factor-alpha is required in the protective immune response against Mycobacterium tuberculosis in mice. Immunity 1995;2:561–572.

40 Bluethmann H: Physiological, immunological, and pathological functions of tumor necrosis factor (TNF) revealed by TNF receptor-deficient mice; in Durum SK, Muegge K (eds): Cytokine Knockouts. Totowa, Humana Press, 1998, pp 69–87.

41 Erickson SL, de Sauvage FJ, Kikly K, Carver Moore K, Pitts Meek S, Gillett N, Sheehan KC, Schreiber RD, Goeddel DV, Moore MW: Decreased sensitivity to tumour-necrosis factor but normal T-cell development in TNF receptor-2-deficient mice. Nature 1994;372:560–563.

42 Wang B, Fujisawa H, Zhuang L, Kondo S, Shivji GM, Kim CS, Mak TW, Sauder DN: Depressed Langerhans cell migration and reduced contact hypersensitivity response in mice lacking TNF receptor p75. J Immunol 1997;159:6148–6155.

43 Lucas R, Juillard P, Decoster E, Redard M, Burger D, Donati Y, Giroud C, Monso Hinard C, De Kesel T, Buurman WA, Moore MW, Dayer JM, Fiers W, Bluethmann H, Grau GE: Crucial role of tumor necrosis factor (TNF) receptor 2 and membrane-bound TNF in experimental cerebral malaria. Eur J Immunol 1997;27:1719–1725.

44 Pfizenmaier K, Wajant H, Grell M: Tumor necrosis factors in 1996. Cytokine Growth Factor Rev 1996;7:271–277.

45 Lazdins JK, Grell M, Walker MR, Woods Cook K, Scheurich P, Pfizenmaier K: Membrane tumor necrosis factor (TNF) induced cooperative signaling of TNFR60 and TNFR80 favors induction of cell death rather than virus production in HIV-infected T cells. J Exp Med 1997;185:81–90.

46   Schmid EF, Binder K, Grell M, Scheurich P, Pfizenmaier K: Both tumor necrosis factor receptors, TNFR60 and TNFR80, are involved in signaling endothelial tissue factor expression by juxtacrine tumor necrosis factor a. Blood 1995;86:1836–1841.

47   Küsters S, Tiegs G, Alexopoulou L, Pasparakis M, Douni E, Kunstle G, Bluethmann H, Wendel A, Pfizenmaier K, Kollias G, Grell M: In vivo evidence for a functional role of both tumor necrosis factor (TNF) receptors and transmembrane TNF in experimental hepatitis. Eur J Immunol 1997; 27:2870–2875.

48   Tartaglia LA, Pennica D, Goeddel DV: Ligand passing: The 75-kDa tumor necrosis factor (TNF) receptor recruits TNF for signaling by the 55-kDa TNF receptor. J Biol Chem 1993;268:18542–18548.

49   Grell M, Wajant H, Zimmermann G, Scheurich P: The type 1 receptor (CD120a) is the high-affinity receptor for soluble tumor necrosis factor. Proc Natl Acad Sci USA 1998;95:570–575.

50   Grell M, Douni E, Wajant H, Lohden M, Clauss M, Maxeiner B, Georgopoulos S, Lesslauer W, Kollias G, Pfizenmaier K: The transmembrane form of tumor necrosis factor is the prime activating ligand of the 80 kDa tumor necrosis factor receptor. Cell 1995;83:793–802.

51   Douni E, Kollias G: A critical role of the p75 tumor necrosis factor receptor (p75TNF-R) in organ inflammation independent of TNF, lymphotoxin α, or the p55TNF-R. J Exp Med 1998;188: 1343–1352.

52   Ledgerwood EC, Prins JB, Bright NA, Johnson DR, Wolfreys K, Pober JS, O'Rahilly S, Bradley JR: Tumor necrosis factor is delivered to mitochondria where a tumor necrosis factor-binding protein is localized. Lab Invest 1999;78:1583–1589.

53   Beutler B, Milsark IW, Cerami AC: Passive immunization against cachectin/tumor necrosis factor protects mice from lethal effect of endotoxin. Science 1985;229:869–871.

54   Tracey KJ, Fong Y, Hesse DG, Manogue KR, Lee AT, Kuo GC, Lowry SF, Cerami A: Anti-cachectin/TNF monoclonal antibodies prevent septic shock during lethal bacteraemia. Nature 1987; 330:662–664.

55   Hinshaw LB, Tekamp Olson P, Chang AC, Lee PA, Taylor FB, Murray CK, Peer GT, Emerson TE, Passey RB, Kuo GC: Survival of primates in LD100 septic shock following therapy with antibody to tumor necrosis factor (TNF a). Circ Shock 1990;30:279–292.

56   Hinshaw LB, Emerson TE, Taylor FB, Chang AC, Duerr M, Peer GT, Flournoy DJ, White GL, Kosanke SD, Murray CK, Xu R, Passey RB, Fournel MA: Lethal *Staphylococcus aureus*-induced shock in primates: Prevention of death with anti-TNF antibody. J Trauma 1992;33:568–573.

57   Silva AT, Bayston KF, Cohen J: Prophylactic and therapeutic effects of a monoclonal antibody to tumor necrosis factor-α in experimental gram-negative shock. J Infect Dis 1990;162:421–427.

58   Tracey KJ, Beutler B, Lowry SF, Merryweather J, Wolpe S, Milsark IW, Hariri RJ, Fahey TJ, Zentella A, Albert JD, Shires GT, Cerami A: Shock and tissue injury induced by recombinant human cachectin. Science 1986;234:470.

59   Waage A, Espevik T: Interleukin 1 potentiates the lethal effect of tumor necrosis factor a/cachectin in mice. J Exp Med 1988;167:1987–1992.

60   Galanos C, Freudenberg MA, Reutter W: Galactosamine-induced sensitization to the lethal effects of endotoxin. Proc Natl Acad Sci USA 1979;76:5939–5943.

61   Tiegs G, Wolter M, Wendel A: Tumor necrosis factor is a terminal mediator in galactosamine/endotoxin-induced hepatitis in mice. Biochem Pharmacol 1989;38:627–631.

62   Bahrami S, Redl H, Leichtfried G, Yu Y, Schlag G: Similar cytokine but different coagulation responses to lipopolysaccharide injection in *D*-galactosamine-sensitized versus nonsensitized rats. Infect Immun 1994;62:99–105.

63   Leist M, Gantner F, Bohlinger I, Germann PG, Tiegs G, Wendel A: Murine hepatocyte apoptosis induced in vitro and in vivo by TNF-α requires transcriptional arrest. J Immunol 1994;153:1778–1788.

64   Michie HR, Manogue KR, Spriggs DR, Revhaug A, O'Dwyer S, Dinarello CA, Cerami A, Wolff SM, Wilmore DW: Detection of circulating tumor necrosis factor after endotoxin administration. N Engl J Med 1988;318:1481–1486.

65   Michie HR, Spriggs DR, Manogue KR, Sherman ML, Revhaug A, O'Dwyer ST, Arthur K, Dinarello CA, Cerami A, Wolff SM: Tumor necrosis factor and endotoxin induce similar metabolic responses in human beings. Surgery 1988;104:280–286.

66 Waage A, Halstensen A, Shalaby R, Brandtzaeg P, Kierulf P, Espevik T: Local production of tumor necrosis factor α, interleukin 1, and interleukin 6 in meningococcal meningitis: Relation to the inflammatory response. J Exp Med 1989;170:1859–1867.

67 Damas P, Reuter A, Gysen P, Demonty J, Lamy M, Franchimont P: Tumor necrosis factor and interleukin-1 serum levels during severe sepsis in humans. Crit Care Med 1989;17:975–978.

68 Marks JD, Marks CB, Luce JM, Montgomery AB, Turner J, Metz CA, Murray JF: Plasma tumor necrosis factor in patients with septic shock: Mortality rate, incidence of adult respiratory distress syndrome, and effects of methylprednisolone administration. Am Rev Respir Dis 1990;141:94–97.

69 Calandra T, Baumgartner JD, Grau GE, Wu MM, Lambert PH, Schellekens J, Verhoef J, Glauser MP: Prognostic values of tumor necrosis factor/cachectin, interleukin-1, interferon-α, and interferon-γ in the serum of patients with septic shock. Swiss-Dutch J5 Immunoglobulin Study Group. J Infect Dis 1990;161:982–987.

70 Müller JM, Ziegler Heitbrock HW, Baeuerle PA: Nuclear factor κ B, a mediator of lipopolysaccharide effects. Immunobiology 1993;187:233–256.

71 Bohrer H, Qiu F, Zimmermann T, Zhang Y, Illmer T, Männel D, Bottiger BW, Stern DM, Waldherr R, Saeger HD, Ziegler R, Bierhaus A, Martin E, Nawroth PP: Role of NFκB in the mortality of sepsis. J Clin Invest 1997;100:972–985.

72 Fisher CJ Jr, Agosti JM, Opal SM, Lowry SF, Balk RA, Sadoff JC, Abraham E, Schein RM, Benjamin E: Treatment of septic shock with the tumor necrosis factor receptor: Fc fusion protein. The Soluble TNF Receptor Sepsis Study Group. N Engl J Med 1996;334:1697–1702.

73 Bone RC, Balk RA, Cerra FB, Dellinger RP, Fein AM, Knaus WA, Schein RM, Sibbald WJ: Definitions for sepsis and organ failure and guidelines for the use of innovative therapies in sepsis. The ACCP/SCCM Consensus Conference Committee. American College of Chest Physicians/Society of Critical Care Medicine. Chest 1992;101:1644–1655.

74 Shenep JL, Mogan KA: Kinetics of endotoxin release during antibiotic therapy for experimental gram-negative bacterial sepsis. J Infect Dis 1984;150:380–388.

75 Danner RL, Elin RJ, Hosseini JM, Wesley RA, Reilly JM, Parillo JE: Endotoxemia in human septic shock. Chest 1991;99:169–175.

76 Holzheimer RG: The significance of endotoxin release in experimental and clinical sepsis in surgical patients: Evidence for antibiotic-induced endotoxin release? Infection 1998;26:77–84.

77 Prins JM, Speelman P, Kuijper EJ, Dankert J, van Deventer SJ: No increase in endotoxin release during antibiotic killing of meningococci. J Antimicrob Chemother 1997;39:13–18.

78 Prins JM, Lauw FN, Derkx BH, Speelman P, Kuijper EJ, Dankert J, van Deventer SJ: Endotoxin release and cytokine production in acute and chronic meningococcaemia. Clin Exp Immunol 1998; 114:215–219.

79 Newcomb D, Bolgos G, Green L, Remick DG: Antibiotic treatment influences outcome in murine sepsis: Mediators of increased morbidity. Shock 1998;10:110–117.

80 Fekade D, Knox K, Hussein K, Melka A, Lalloo DG, Coxon RE, Warrell DA: Prevention of Jarisch-Herxheimer reactions by treatment with antibodies against tumor necrosis factor a. N Engl J Med 1996;335:311–315.

81 Vidal V, Scragg IG, Cutler SJ, Rockett KA, Fekade D, Warrell DA, Wright DJ, Kwiatkowski D: Variable major lipoprotein is a principal factor of louse-borne relapsing fever. Nat Med 1998;4: 1416–1420.

82 Wichterman KA, Baue AE, Chaudry IH: Sepsis and septic shock: A review of laboratory models and a proposal. J Surg Res 1980;29:189–201.

83 Wiktor-Jedrzejczak W, Dzwigala B, Szperl M, Maruszynski M, Urbanowska E, Szwech P: Colony-stimulating factor 1-dependent resident macrophages play a regulatory role in fighting *Escherichia coli* fecal peritonitis. Infect Immun 1996;64:1577–1581.

84 O'Brien AD, Rosenstreich DL, Scher I, Campbell GH, MacDermott RP, Formal SB: Genetic control of susceptibility to *Salmonella typhimurium* in mice: Role of the LPS gene. J Immunol 1980; 124:20–24.

85 Cross AS, Sadoff JC, Kelly N, Bernton E, Gemski P: Pretreatment with recombinant murine tumor necrosis factor a/cachectin and murine interleukin 1 a protects mice from lethal bacterial infection. J Exp Med 1989;169:2021–2027.

86 Poltorak A, He X, Smirnova I, Liu MY, Huffel CV, Du X, Birdwell D, Alejos E, Silva M, Galanos C, Freudenberg M, Ricciardi-Castagnoli P, Layton B, Beutler B: Defective LPS signaling in C3H/HeJ and C57BL/10ScCr mice: Mutations in Tlr4 gene. Science 1998;282:2085.

87 Jack RS, Fan X, Bernheiden M, Rune G, Ehlers M, Weber A, Kirsch G, Mentel R, Furll B, Freudenberg M, Schmitz G, Stelter F, Schütt C: Lipopolysaccharide-binding protein is required to combat a murine gram-negative bacterial infection. Nature 1997;389:742–745.

88 Haziot A, Ferrero E, Kontgen F, Hijiya N, Yamamoto S, Silver J, Stewart CL, Goyert SM: Resistance to endotoxin shock and reduced dissemination of gram-negative bacteria in CD14-deficient mice. Immunity 1996;4:407–414.

89 Wright SD: CD14 and innate recognition of bacteria. J Immunol 1995;155:6–8.

90 Wang SC, Klein RD, Wahl WL, Alarcon WH, Garg RJ, Remick DG, Su GL: Tissue coexpression of LBP and CD14 mRNA in a mouse model of sepsis. J Surg Res 1998;76:67–73.

91 Mercer Jones MA, Heinzelmann M, Peyton JC, Wickel DJ, Cook M, Cheadle WG: The pulmonary inflammatory response to experimental fecal peritonitis: Relative roles of tumor necrosis factor-α and endotoxin. Inflammation 1997;21:401–417.

92 Lucas R, Echtenacher B, Sablon E, Juillard P, Magez S, Lou J, Donati Y, Bosman F, Van de Voorde A, Fransen L, Männel DN, Grau GE, De Baetselier P: Generation of a mouse tumor necrosis factor mutant with antiperitonitis and desensitization activities comparable to those of the wild type but with reduced systemic toxicity. Infect Immun 1997;65:2006–2010.

93 Echtenacher B, Falk W, Männel DN, Krammer PH: Requirement of endogenous tumor necrosis factor/cachectin for recovery from experimental peritonitis. J Immunol 1990;145:3762–3766.

94 Havell EA: Evidence that tumor necrosis factor has an important role in antibacterial resistance. J Immunol 1989;143:2894–2899.

95 Echtenacher B, Hültner L, Männel DN: Cellular and molecular mechanisms of TNF protection in septic peritonitis. J Inflamm 1995;47:85–89.

96 Heumann D, Le Roy D, Zanetti G, Eugster HP, Ryffel B, Hahne M, Tschopp J, Glauser MP: Contribution of TNF/TNF receptor and of Fas ligand to toxicity in murine models of endotoxemia and bacterial peritonitis. J Inflamm 1995;47:173–179.

97 Zhan Y, Cheers C: Control of IL-12 and IFN-γ production in response to liver or dead bacteria by TNF and other factors. J Immunol 1998;161:1447–1453.

98 Roake JA, Rao AS, Morris PJ, Larsen CP, Hankins DF, Austyn JM: Dendritic cell loss from nonlymphoid tissues after systemic administration of lipopolysaccharide, tumor necrosis factor, and interleukin 1. J Exp Med 1995;181:2237–2247.

99 Cumberbatch M, Dearman RJ, Kimber I: Langerhans cells require signals from both tumour necrosis factor-α and interleukin-1 β for migration. Immunology 1997;92:388–395.

100 Vacheron F, Rudent A, Perin S, Labarre C, Quero AM, Guenounou M: Production of interleukin 1 and tumour necrosis factor activities in bronchoalveolar washings following infection of mice by influenza virus. J Gen Virol 1990;71:477–479.

101 Waage A, Halstensen A, Espevik T: Association between tumour necrosis factor in serum and fatal outcome in patients with meningococcal disease. Lancet 1987;i:355–357.

102 Louie A, Baltch AL, Smith RP, Franke MA, Ritz WJ, Singh JK, Gordon MA: Tumor necrosis factor α has a protective role in a murine model of systemic candidiasis. Infect Immun 1994;62:2761–2772.

103 Hultgren O, Eugster HP, Sedgwick JD, Körner H, Tarkowski A: TNF/lymphotoxin-alpha double-mutant mice resist septic arthritis but display increased mortality in response to *Staphylococcus aureus*. J Immunol 1998;161:5937–5942.

104 Männel DN, Echtenacher B: The role of tumor necrosis factor (TNF) in the host defense mechanisms against infection; in Baumgartner JD, Calandra T, Carlet J (eds): Mediators of Sepsis from Pathophysiology to Therapeutic Approaches. Paris, Flammarion, 1993, pp 63–69.

105 Gordon JR, Galli SJ: Mast cells as a source of both preformed and immunologically inducible TNF-α/cachectin. Nature 1990;346:274–276.

106 Gordon JR, Galli SJ: Release of both preformed and newly synthesized tumor necrosis factor a (TNF-α)/cachectin by mouse mast cells stimulated via the FcϵRI: A mechanism for the sustained action of mast cell-derived TNF-α during IgE-dependent biological responses. J Exp Med 1991; 174:103–107.

Männel/Echtenacher

107 Echtenacher B, Männel DN, Hültner L: Critical protective role of mast cells in a model of acute septic peritonitis. Nature 1996;381:75–77.

108 Malaviya R, Ikeda T, Ross E, Abraham SN: Mast cell modulation of neutrophil influx and bacterial clearance at sites of infection through TNF-α. Nature 1996;381:77–80.

109 Klimpel GR, Langley KE, Wypych J, Abrams JS, Chopra AK, Niesel DW: A role for stem cell factor (SCF): c-Kit interaction(s) in the intestinal tract response to *Salmonella typhimurium* infection. J Exp Med 1996;184:271–276.

110 Malaviya R, Abraham SN: Clinical implications of mast cell-bacteria interaction. J Mol Med 1998; 76:617–623.

111 Abraham SN, Malaviya R: Mast cells in infection and immunity. Infect Immun 1997;65:3501–3508.

112 Abramson JS, Dahl MV, Walsh G, Blumenthal MN, Douglas SD, Quie PG: Antistaphylococcal IgE in patients with atopic dermatitis. J Am Acad Dermatol 1982;7:105–110.

113 Aceti A, Celestino D, Caferro M, Casale V, Citarda F, Conti EM, Grassi A, Grilli A, Pennica A, Sciarretta F: Basophil-bound and serum immunoglobulin E directed against *Helicobacter pylori* in patients with chronic gastritis. Gastroenterology 1991;101:131–137.

114 Leung DY, Harbeck R, Bina P, Reiser RF, Yang E, Norris DA, Hanifin JM, Sampson HA: Presence of IgE antibodies to staphylococcal exotoxins on the skin of patients with atopic dermatitis. Evidence for a new group of allergens. J Clin Invest 1993;92:1374–1380.

115 Zhang Y, Ramos BF, Jakschik BA: Neutrophil recruitment by tumor necrosis factor from mast cells in immune complex peritonitis. Science 1992;258:1957–1959.

116 Miyajima I, Dombrowicz D, Martin TR, Ravetch JV, Kinet JP, Galli SJ: Systemic anaphylaxis in the mouse can be mediated largely through IgG1 and FcγRIII: Assessment of the cardiopulmonary changes, mast cell degranulation, and death associated with active or IgE- or IgG1-dependent passive anaphylaxis. J Clin Invest 1997;99:901–914.

117 Dombrowicz D, Flamand V, Miyajima I, Ravetch JV, Galli SJ, Kinet JP: Absence of FcεRI a chain results in upregulation of FcγRIII-dependent mast cell degranulation and anaphylaxis. Evidence of competition between FcεRI and FcγRIII for limiting amounts of FcR β and γ chains. J Clin Invest 1997;99:915–925.

118 Baorto DM, Gao Z, Malaviya R, Dustin ML, van der Merwe A, Lublin DM, Abraham SN: Survival of FimH-expressing enterobacteria in macrophages relies on glycolipid traffic. Nature 1997; 389:636–639.

119 Malaviya R, Ross E, Jakschik BA, Abraham SN: Mast cell degranulation induced by type 1 fimbriated *Escherichia coli* in mice. J Clin Invest 1994;93:1645–1653.

120 Sher A, Hein A, Moser G, Caulfield JP: Complement receptors promote the phagocytosis of bacteria by rat peritoneal mast cells. Lab Invest 1979;41:490–499.

121 Sher A, McIntyre SL: Receptors for C3 on rat peritoneal mast cells. J Immunol 1977;119:722–725.

122 Sher A: Complement-dependent adherence of mast cells to schistosomula. Nature 1976;263:334–336.

123 Ramos BF, Zhang Y, Jakschik BA: Neutrophil elicitation in the reverse passive Arthus reaction: Complement-dependent and -independent mast cell involvement. J Immunol 1994;152:1380–1384.

124 Fischer MB, Prodeus AP, Nicholson Weller A, Ma M, Murrow J, Reid RR, Warren HB, Lage AL, Moore FD Jr, Rosen FS, Carroll MC: Increased susceptibility to endotoxin shock in complement C3- and C4-deficient mice is corrected by C1 inhibitor replacement. J Immunol 1997;159:976–982.

125 Wessels MR, Butko P, Ma M, Warren HB, Lage AL, Carroll MC: Studies of group B streptococcal infection in mice deficient in complement component C3 or C4 demonstrate an essential role for complement in both innate and acquired immunity. Proc Natl Acad Sci USA 1995;92:11490–11494.

126 Prodeus AP, Zhou X, Maurer M, Galli SJ, Carroll MC: Impaired mast cell-dependent natural immunity in complement C3-deficient mice. Nature 1997;390:172–175.

127 Höpken UE, Lu B, Gerard NP, Gerard C: The C5a chemoattractant receptor mediates mucosal defence to infection. Nature 1996;383:86–89.

128 Höpken UE, Lu B, Gerard NP, Gerard C: Impaired inflammatory responses in the reverse arthus reaction through genetic deletion of the C5a receptor. J Exp Med 1997;186:749–756.

129 Boes M, Prodeus AP, Schmidt T, Carroll MC, Chen J: A critical role of natural immunoglobulin M in immediate defense against systemic bacterial infection. J Exp Med 1998;188:2381–2386.

130 Ansel JC, Brown JR, Payan DG, Brown MA: Substance P selectively activates TNF-α gene expression in murine mast cells. J Immunol 1993;150:4478–4485.

131 Castagliuolo I, Riegler M, Pasha A, Nikulasson S, Lu B, Gerard C, Gerard NP, Pothoulakis C: Neurokinin-1 (NK-1) receptor is required in *Clostridium difficile*-induced enteritis. J Clin Invest 1998;101:1547–1550.

132 Dickerson C, Undem B, Bullock B, Winchurch RA: Neuropeptide regulation of proinflammatory cytokine responses. J Leukoc Biol 1998;63:602–605.

133 Leal Berumen I, Conlon P, Marshall JS: IL-6 production by rat peritoneal mast cells is not necessarily preceded by histamine release and can be induced by bacterial lipopolysaccharide. J Immunol 1994; 152:5468–5476.

134 Sayers TJ, Wiltrout TA, Bull CA, Denn AC, Pilaro AM, Lokesh B: Effect of cytokines on polymorphonuclear neutrophil infiltration in the mouse: Prostaglandin- and leukotriene-independent induction of infiltration by IL-1 and tumor necrosis factor. J Immunol 1988;141:1670–1677.

135 Wershil BK, Wang ZS, Gordon JR, Galli SJ: Recruitment of neutrophils during IgE-dependent cutaneous late phase reactions in the mouse is mast cell-dependent: Partial inhibition of the reaction with antiserum against tumor necrosis factor-α. J Clin Invest 1991;87:446–453.

136 Chaka W, Verheul AF, Vaishnav VV, Cherniak R, Scharringa J, Verhoef J, Snippe H, Hoepelman AI: Induction of TNF-a in human peripheral blood mononuclear cells by the mannoprotein of *Cryptococcus neoformans* involves human mannose binding protein. J Immunol 1997;159:2979–2985.

137 Echtenacher B, Falk W, Männel DN, Krammer PH: Survival from cecal ligation and puncture and the formation of fibrous adhesions in the peritoneal cavity depend on endogenous tumor necrosis factor; in Faist E, Meakins JL, Schildberg FW (eds): Host Defense Dysfunction in Trauma, Shock and Sepsis. Heidelberg, Springer, 1993, pp 755–758.

138 Gibson FC, Onderdonk AB, Kasper DL, Tzianabos AO: Cellular mechanism of intraabdominal abscess formation by *Bacteroides fragilis*. J Immunol 1998;160:5000–5006.

139 Kaidi AA, Gurchumelidze T, Nazzal M, Figert P, Vanterpool C, Silva Y: Tumor necrosis factor-α: A marker for peritoneal adhesion formation. J Surg Res 1995;58:516–518.

140 van der Poll T, Buller HR, ten Cate H, Wortel CH, Bauer KA, van Deventer SJ, Hack CE, Sauerwein HP, Rosenberg RD, ten Cate JW: Activation of coagulation after administration of tumor necrosis factor to normal subjects. N Engl J Med 1990;322:1622–1627.

141 Levi M, van der Poll T, ten Cate H, van Deventer SJ: The cytokine-mediated imbalance between coagulant and anticoagulant mechanisms in sepsis and endotoxaemia. Eur J Clin Invest 1997;27: 3–9.

142 van der Poll T, Levi M, Van Deventer SJH, ten Cate H, Haagmans BL, Biemond BJ, Büller HR, Hack CE, ten Cate JW: Differential effects of anti-tumor necrosis factor monoclonal antibodies on systemic inflammatory responses in experimental endotoxemia in chimpanzees. Blood 1994;83: 446–451.

143 DeLa Cadena RA, Majluf Cruz A, Stadnicki A, Tropea M, Reda D, Agosti JM, Colman RW, Suffredini AF: Recombinant tumor necrosis factor receptor p75 fusion protein (TNFR:Fc) alters endotoxin-induced activation of the kinin, fibrinolytic, and coagulation systems in normal humans. Thromb Haemost 1998;80:114–118.

144 van der Poll T, Jansen PM, Van Zee KJ, Hack CE, Oldenburg HA, Loetscher H, Lesslauer W, Lowry SF, Moldawer LL: Pretreatment with a 55-kDa tumor necrosis factor receptor-immunoglobulin fusion protein attenuates activation of coagulation, but not of fibrinolysis, during lethal bacteremia in baboons. J Infect Dis 1997;176:296–299.

145 Carr C, Bild GS, Chang AC, Peer GT, Palmier MO, Frazier RB, Gustafson ME, Wun TC, Creasey AA, Hinshaw LB: Recombinant *E. coli*-derived tissue factor pathway inhibitor reduces coagulopathic and lethal effects in the baboon gram-negative model of septic shock. Circ Shock 1994;44:126–137.

146 Creasey AA, Chang AC, Feigen L, Wun TC, Taylor FB, Hinshaw LB: Tissue factor pathway inhibitor reduces mortality from *Escherichia coli* septic shock. J Clin Invest 1993;91:2850–2856.

147 Goldfarb RD, Glock D, Johnson K, Creasey AA, Carr C, McCarthy RJ, Matushek M, Akhter I, Trenholme G, Parrillo JE: Randomized, blinded, placebo-controlled trial of tissue factor pathway inhibitor in porcine septic shock. Shock 1998;10:258–264.

---

148 Guerder S, Picarella DE, Linsley PS, Flavell RA: Costimulator B7-1 confers antigen-presenting cell function to parenchymal tissue and in conjunction with tumor necrosis factor a leads to autoimmunity in transgenic mice. Proc Natl Acad Sci USA 1994;91:5138–5142.

149 Picarella DE, Kratz A, Li CB, Ruddle NH, Flavell RA: Transgenic tumor necrosis factor (TNF)-α production in pancreatic islets leads to insulitis, not diabetes: Distinct patterns of inflammation in TNF-α and TNF-β transgenic mice. J Immunol 1993;150:4136–4150.

150 Ruddle NH, Bergman CM, McGrath KM, Lingenheld EG, Grunnet ML, Padula SJ, Clark RB: An antibody to lymphotoxin and tumor necrosis factor prevents transfer of experimental allergic encephalomyelitis. J Exp Med 1990;172:1193–1200.

151 Liu J, Marino MW, Wong G, Grail D, Dunn A, Bettadapura J, Slavin AJ, Old L, Bernard CC: TNF is a potent anti-inflammatory cytokine in autoimmune-mediated demyelination. Nat Med 1998;4:78–83.

152 Bickston SJ, Cominelli F: Treatment of Crohn's disease at the turn of the century. N Engl J Med 1998;339:401–402.

153 O'Dell JR: Anticytokine therapy – A new era in the treatment of rheumatoid arthritis? N Engl J Med 1999;340:310–312.

154 van Oosten BW, Barkhof F, Truyen L, Boringa JB, Bertelsmann FW, von Blomberg BM, Woody JN, Hartung HP, Polman CH: Increased MRI activity and immune activation in two multiple sclerosis patients treated with the monoclonal anti-tumor necrosis factor antibody cA2. Neurology 1996;47:1531–1534.

Daniela N. Männel, Institut für Pathologie, Tumorimmunologie, Universität Regensburg,
D–93042 Regensburg (Germany)
Tel. +49 941 944 6622, Fax +49 941 944 6602, E-Mail daniela.maennel@klinik.uni-regensburg.de

Jack RS (ed): CD14 in the Inflammatory Response.
Chem Immunol. Basel, Karger, 2000, vol 74, pp 162–177

....

Clinical Aspects: From Systemic Inflammation to 'Immunoparalysis'

Hans-Dieter Volk[a]*, Petra Reinke*[b]*, Wolf-Dietrich Döcke*[a]

[a] Institute of Medical Immunology, Charité-Campus Mitte, and
[b] Department of Nephrology and Internal Intensive Care, Charité-Campus Virchow,
Humboldt University Berlin, Germany

Antimicrobial defense against pathogens plays an important role in the protection of the body's integrity. We can distinguish three levels of antimicrobial defense: (1) the static natural barriers (skin, mucosa, pH, natural flora in intestine and vagina); (2) the inducible acute inflammatory system (complement, mast cells, phagocytes, NK cells, interferons, lysozyme, acute-phase response, etc.), and (3) the adaptive immune response with memory (lymphocytes, antigen-presenting cells).

Disturbances of the natural barriers, as seen in ICU patients, dramatically increase the risk of invasive infections. In fact, in ICU patients infections are one of the main clinical complications and they may develop into sepsis which is associated with a very high mortality. Over the last two decades our understanding of the pathophysiology of sepsis has progressed considerably. It is now clear that an overwhelming systemic inflammatory response to invasion by bacteria and fungi and/or their microbial toxins may be involved in the pathogenesis of SIRS, sepsis and multiple-organ failure. Since TNF and IL-1 can mimic sepsis and septic shock in animal models, several recent clinical trials have focussed on neutralisation of these inflammatory mediators. These trials, however, have had very disappointing results.

Several reasons for the failure of the anti-inflammatory approach in sepsis have been discussed [1, 2]. From the immunological point of view we had to learn that the accepted model of sepsis pathogenesis in which invasion of bacteria and/or toxins was followed by systemic release of proinflammatory cytokines and that this led then to shock and to multiple-organ failure was oversimplified [3].

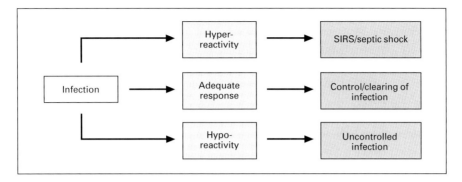

Fig. 1. The immune responsiveness determines the clinical outcome of invasive infections.

The Beneficial Effects of Inflammation in Severe Infections

During the 1990s, many scientists (and companies) regarded TNF and IL-1 as the 'bad guys' in sepsis despite the fact that important experimental and clinical data contradicting this viewpoint were available. For example TNF targeting therapy was shown to be successful in preventing shock following a bolus i.v. or i.p. application of endotoxin (LPS) or bacteria in animal models. However, it showed no benefit or was even frankly deleterious in the cecal ligation and puncture (CLP) peritonitis model which is closer to the clinical situation [4]. Moreover, mice which are genetically deficient in the production or response to TNF are relatively resistant to LPS-mediated injury but they are deficient in the control of an infection with living bacteria [5]. In case of pathogens this deficiency was associated with death even at low dose infections. Similar observations were made with LBP k.o. mice [6]. On the other hand, mice deficient in anti-inflammatory cytokines, like TGF-β or IL-10, are very sensitive to LPS or bacterial-induced shock as the result of an uncontrolled inflammatory response.

From all of this and from the disappointing results of the clinical trials of anti-inflammatory therapies, we learnt that our models were too simple. A bolus injection of LPS or bacteria does not adequately reflect the complex host-pathogen interactions that take place during sepsis. These data suggest that a well-balanced inflammatory response is important to control bacterial infections. Too much inflammation in response to infection resulting in hyper-inflammation can be as lethal as too little inflammation resulting in an uncontrolled infection (fig. 1).

We well understand the mechanisms of shock, multiple-organ failure and death induced by hyperinflammation, but further studies are necessary to learn more about the pathogenetic factors involved in the 'hypoinflammatory' infection models – for example, we do not know why the LBP –/– mice die from low-dose *Salmonella* infection. Is the host's inflammatory system involved?

Clinical data also challenge the simple concept of septic death 'by too much inflammation'. Proinflammatory and anti-inflammatory cytokines have been detected in serum and body fluids of septic patients [7]. However, attempts to build up a picture of what is going on from these data are bedevilled by the complexity of the networks involved and the short half-life of the participating cytokines. In addition, the cytokine concentration in the serum does not automatically reflect its tissue concentration, in particular its concentration at the site of infection. Because of this it is not terribly surprising that in some studies sepsis outcome was found to be associated with an increase in serum levels of proinflammatory cytokines while in others the association was to increased serum levels of anti-inflammatory cytokines.

Several groups including ours have reported deactivation of monocytes and granulocytes in septic patients, particularly during later stages of disease [3]. In contrast to the rapid and erratic fluctuations in serum cytokine concentrations this monocyte phenotype is remarkably stable. Whereas cytokines have half-lifes in plasma measured in minutes, monocytes and granulocytes after leaving the bone marrow, have a half-life in the peripheral blood of about 24 h. Monocytes migrate into the tissues where they differentiate into different subtypes of macrophages, but the granulocytes are short-lived and can only go into inflammatory tissues. These properties make analysis of monocyte and granulocyte function suitable for daily monitoring.

In prolonged sepsis we demonstrated a state of the inflammatory system in which monocytic TNF secretion capacity as well as HLA-DR antigen expression and antigen-presenting capacity are all severely depressed. However, the capacity to produce IL-1ra and, at least temporarily, also IL-10 is largely preserved. This switch to predominant anti-inflammatory mediator release was associated with a high risk of fatal outcome from persistent, in part opportunistic, infections which lead to multiple-organ failure. In fact, none of the patients we have monitored survived unless their monocytes recovered proinflammatory function and MHC-class II expression. We initially observed this phenomenon in septic transplant patients. In those patients with reduced monocyte proinflammatory functions, a rapid reduction of immunosuppression usually did not lead to acute graft rejection. In contrast, septic transplant patients with normal monocytic phenotype and function rapidly rejected their graft if immunosuppression was reduced. Clearly, deactivation of monocytic proinflammatory functions reflects a generalised in vivo deactivation of the entire cell-mediated innate immune system. Therefore, we called this phenomenon 'immunoparalysis' [8, 9].

What Are the Mechanisms of Monocyte Deactivation?

Monocytes are a crucial component of resistance to infection. They phagocytose and kill pathogenic microorganisms, neutralize toxins originating from pathogens and as antigen-presenting cells they are an important link between the innate resistance system and the highly specialized adaptive immune response. All of these functions are disturbed in monocytes from patients with 'immunoparalysis'.

In searching for factors which trigger 'immunoparalysis', we noticed that its occurrence was not dependent on specific pathogens or toxins. It was striking, however, that almost every case of *Candida* sepsis (usually appearing in combination with bacterial sepsis) was associated with 'immunoparalysis' though by no means all immunoparalysis patients had a *Candida* infection. The *Candida* sepsis can therefore be regarded as an opportunistic infection which results from rather than causes the immunodepression. Another indication that 'immunoparalysis' is not the direct result of pathogenic action is the fact that we and others have observed monocyte deactivation in some patients after sterile injury (burns, trauma, major surgery) or in the course of high-dose immunosuppression (induction or anti-rejection therapy) [7–9]. Monocyte phenotype and function are closely associated with the 'net' immunosuppression of cell-mediated immunity, particularly of the type 1 T-cell response. The proinflammatory capacity of monocytes (secretion of TNF, IL-1, IL-12, etc.) as well as their antigen-presenting capacity (expression of surface HLA-DR, HLA-DP, HLA-DQ, CD80/86) is positively regulated by immunostimulatory cytokines such as interferon-γ (IFN-γ) or GM-CSF and negatively influenced by several factors including IL-10, TGF-β, prostaglandins, catecholamines and apoptotic material.

Following high-dose immunosuppression in transplant patients using steroid bolus, OKT 3 mAb or ATG, we have seen a temporary decrease both in monocytic HLA-DR expression and in the capacity to generate TNF ex vivo. The reason for this phenomenon may be the lack of stimulatory cytokines such as IFN-γ which is strongly downregulated by immunosuppression. In addition, immunosuppression will induce inhibitory factors since steroids block cytokine action on monocytes and other cells while cyclosporine upregulates TGF-β secretion. In some patients monocyte deactivation even reaches the level of 'immunoparalysis' – a situation which is associated with an increased risk to develop infectious complications within the next few weeks. Indeed after 2 days of immunosuppression-induced 'immunoparalysis' bacterial and/or fungal infections were seen in about 30% of patients vs. 4% in patients without 'immunoparalysis'. The longer the 'immunoparalysis' persists the greater the incidence of infections [9]. In general, the immunocompetence rapidly recovers following reduction of immunosuppression.

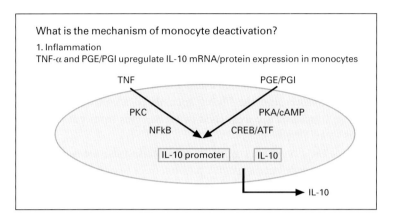

Fig. 2. Mechanisms of monocyte deactivation. I. Inflammation.

But what are the mechanisms of 'immunoparalysis' in nontransplanted ICU patients? Following in vitro LPS stimulation of murine or human blood leukocytes, TNF secretion peaks about 2 h later whereas peak secretion of IL-10 needs more than 14 h. TNF by itself can induce IL-10 mRNA and in combination with other stimuli such as low-dose LPS or prostaglandins can also induce IL-10 protein synthesis. Moreover, if the secreted IL-10 is inactivated by neutralizing antibodies then the secretion of IL-10 was also inhibited suggesting a negative feedback regulation. A high concentration of TNF leads to hyperinflammation but at the same time induces IL-10 which will tend to downregulate the inflammatory reaction. In fact, in vitro treatment of monocytes with IL-10 induces monocyte deactivation, and neutralisation of IL-10 blocks LPS desensitisation, a phenomenon in which monocytes are deactivated by a prior exposure to a low dose of LPS [3, 10]. These data suggest that inflammation by itself may trigger its own downregulation via the NFkB pathway induced by TNF and via the cAMP pathway induced by prostaglandins (fig. 2). The monocyte deactivation frequently observed in patients surviving a septic shock (late sepsis) may be related to this phenomenon. However, this seems to be not the only mechanism. Uptake of apoptotic material by monocytes also downregulates TNF secretion capacity and upregulates IL-10 secretion (fig. 3). The molecular mechanism is not fully understood but the CD36 receptor seems to be involved. A massive induction of apoptosis in organ failure (e.g. liver failure) may contribute to the phenomenon of 'immunoparalysis' in septic and trauma patients.

Loss of monocytic HLA-DR expression and diminished ex vivo TNF secretion capacity have also been observed in many noninfected patients after

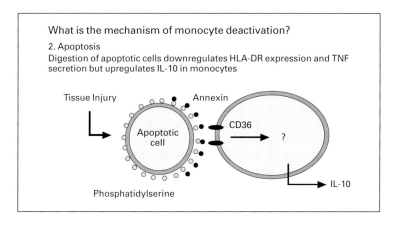

Fig. 3. Mechanisms of monocyte deactivation. II. Apoptosis.

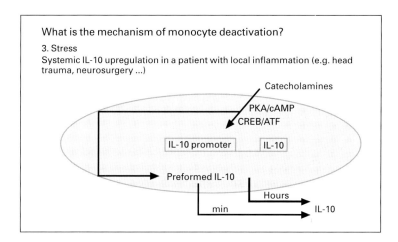

Fig. 4. Mechanisms of monocyte deactivation. III. Stress reaction.

trauma, burn or major surgery. How is this to be explained? It appears that the activation of the stress response following local or systemic inflammation plays an important role in downregulating systemic immune responsiveness. The regulatory role of activation of the HPA axis resulting in corticosteroid release is long established. Very recently we have shown that activation of the catecholamine axis is also involved in a major way in monocyte deactivation following stressful events. The molecular mechanisms of this phenomenon have been investigated and again cAMP-dependent upregulation of IL-10 turns out to play a key role (fig. 4) [11]. Experimental studies in rats confirmed the interpretation of the clinical observations. In these experiments brain trauma was induced either by an intraventricular infusion of proinflammatory cytokines simulating head trauma, or by a local injury to the brain induced by an intraventricular balloon which caused an increase of intracranial pressure. Neither procedure was associated with systemic release of the proinflammatory cytokines TNF, IL-1, IFNγ or IL-6 but both protocols resulted in a rapid systemic release of IL-10, and this could be blocked by parallel application of β-adrenergic receptor blockers [11].

Moreover, this axis seems to be also involved in the regulation of IL-10 released during systemic inflammation. In contrast to the situation in vitro, in both humans and mice TNF and IL-10 are secreted in parallel in vivo within 1 h following LPS challenge suggesting the involvement of TNF-independent mechanisms for IL-10 regulation. Using RT-PCR analyses we found a very high TNF and IL-10 mRNA expression in mouse or rat livers 1 hour after LPS injection while peripheral blood leukocytes expressed TNF only. Eight hours later, IL-10 mRNA was also detectable in blood leukocytes and the protein was, in contrast to TNF, still detectable in the plasma. Both late IL-10 mRNA and IL-10 protein expression were inhibited by the application of a neutralizing anti-TNF Ab in parallel with LPS. In contrast, the early IL-10 synthesis was not reduced following blocking of TNF. However, when β-adrenergic receptor blockers were given along with the LPS not only was early IL-10 release strongly inhibited but the early TNF release was increased [Volk et al., in prep.].

The patient's predisposition is a further factor which clearly has a major impact on the balance between inflammation and anti-inflammation. Thus, we and others have shown that patients with an advanced stage tumour have – already preoperatively – a diminished inflammatory response which is further reduced by the trauma associated with the surgical operation to remove the tumour. Moreover, the capacity to produce cytokines in response to a trigger differs from one individual to another. Interestingly, recent data suggest that the capacity to produce proinflammatory or anti-inflammatory cytokines may be genetically determined. Allelic polymorphisms have been described which are associated with low or high IL-10 and TNF production, respectively [12].

Recently, it was reported that in febrile patients, most of whom had a bacterial infection, a high IL-10/TNF ratio on admission was associated with increased mortality [13]. This fits well to the hypothesis that IL-10 plays a key role in downregulation of the inflammatory system. In fact, septic patients with 'immunoparalysis' expressed low levels of TNF mRNA (table 1) but high levels of IL-10 mRNA (table 2) while septic patients without 'immunoparalysis' showed the opposite picture and had a better outcome [Syrbe et al., submitted].

Table 1. Low TNF-α gene expression in septic patients with immunoparalysis

Detection of TNF mRNA in PBMC by PCR technique (competitive PCR)

Septic patients:	+ +	+	–
Immunoparalysis (HLA-DR+ Mo < 30%)	1	1	16
No immunoparalysis (HLA-DR+ Mo > 45%)	16	10	4
Borderline group (HLA-DR+ Mo 30–45%)	2	3	4
Healthy donors	0	8	8

mRNA expression was measured using semiquantitative RT-PCR (competitive RT-PCR) from reverse-transcribed total RNA of peripheral blood mononuclear cells.

+ + = > 1,000 AU; + = > 0–1,000 AU; – = not detectable (detection limit
about 100 copies).

Table 2. Elevated IL-10 gene expression in septic patients with immunoparalysis

Detection of IL-10 mRNA in PBMC by PCR technique (competitive PCR)

Septic patients:	+ +	+	–
Immunoparalysis (HLA-DR+ Mo < 30%)	8	6	4
No Immunoparalysis (HLA-DR+ Mo > 45%)	4	8	18
Borderline group (HLA-DR+ Mo 30–45%)	4	3	2
Healthy donors	0	0	11

mRNA expression was measured using semiquantitative RT-PCR (competitive RT-PCR) from reverse-transcribed total RNA of peripheral blood mononuclear cells.

+ + = > 100 AU; + = > 0–100 AU, – = not detectable (detection limit
about 100 copies).

Is IL-10 the Only Mediator of Monocyte Deactivation in ICU Patients?

Plasma samples from patients with monocyte deactivation following major
surgery or septic patients but not those from adequate controls inhibited the
monocytic HLA-DR expression and ex vivo TNF secretion of indicator cells
from healthy donors. In most samples elevated IL-10 levels were detectable,
and the inhibitory activity was heat labile and nondialyzable. However, neutralizing IL-10 mAb showed variable effects in that it abolished the inhibitory

activity of the plasma samples by 20–100%. This suggests that, at least in some samples, additional immunosuppressive factors are present [Syrbe et al., submitted].

Moreover, we recently performed a clinical trial in which IL-10 was used for treatment of psoriasis. The patients received up to 12 µg/kg bw/day IL-10 s.c. for several weeks resulting in plasma IL-10 levels comparable with those seen in many ICU patients (20–100 pg/ml). As expected, monocytic HLA-DR expression, antigen-presenting activity, ex vivo TNF/IL-12 secretion capacity, and the Th1/Th2 cytokine ratio were significantly inhibited [14]. However, we never observed such a strong downregulation of monocyte function as is seen in ICU patients. Again this suggests that IL-10 is an important player in monocyte deactivation – but not the only one.

What Is the Cellular Source of the High IL-6 Plasma Levels in Septic Patients with 'Immunoparalysis'?

Despite the monocyte deactivation (including deficient TNF-α/IL-6 secretion), many septic patients with 'immunoparalysis' have moderately increased TNF and highly elevated IL-6 plasma levels – how is this explainable?

First, monocyte deactivation is not automatically identical with tissue macrophage deactivation. However, the longer 'immunoparalysis' persist the more macrophages will be replaced by 'deactivated' monocytes. This is in line with the observation that it is only long-lasting, rather than transient, 'immunoparalysis' which is associated with poor outcome.

Secondly, even in case of monocyte deactivation we never observed complete blocking of TNF/IL-6 secretion which might explain the low-level cytokine levels in patients with persistent infection. Thirdly, IL-6 is not only produced by immune cells. It is also secreted by non-immune cells including endothelial cells, fibroblasts and keratinocytes which secrete IL-6 in response to inflammatory cytokines secreted by macrophages, to direct contact with LPS via sCD14 or to hypoxia. In other words, IL-6 is a marker of tissue injury rather than being a specific marker for the activation of the monocyte/macrophage system.

How Can We Restore the Immune Responsiveness?

If 'immunoparalysis' is not just an epiphenomenon but is in fact causally related to an increased incidence of infectious complications and to a poor prognosis in septic patients, then promoting recovery from immunodeficiency

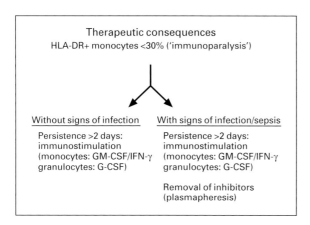

Fig. 5. Putative therapeutic approaches in cases of 'immunoparalysis'.

should be a novel therapeutic approach. The first evidences that this may be a feasible concept came from transplant patients. In patients undergoing high-dose immunosuppression as induction or anti-rejection therapy, we monitor monocyte function two to three times per week. HLA-DR expression is determined by flow cytometry and ex vivo TNF secretion by a semiautomatic ELISA. 'Overimmunosuppression', defined as <30% HLA-DR + monocytes and strongly depressed TNF secretion, is associated with an increased risk of infectious complications within the next 2–3 weeks. In these cases, we reduce the immunosuppression carefully and additionally monitor procalcitonin and IL-6 plasma levels in order to quickly detect incipient systemic infections. This decreases the incidence of severe infections in our transplant population.

If 'immunoparalysis' is observed in transplant patients with established infections, we immediately taper down immunosuppression in order to prevent death by sepsis. With one exception, we have never seen acute graft rejection as a result of reducing immunosuppression in patients with 'immunoparalysis' which underscores the general nature of the immunodeficiency in these patients. In sharp contrast, reduction of immunosuppression in infectious transplant patients without 'immunoparalysis' is frequently followed by acute rejection [8, 9].

In summary, the transplant data suggest that the recovery of an inflammatory response is important for the control of bacterial/fungal infections and that it is not harmful even in septic patients.

These observations were the rationale for a similar approach in nontransplant septic patients (fig. 5). In vitro, ex vivo and in animal models, we first showed that the monocytic activators IFN-γ and GM-CSF were able to reverse

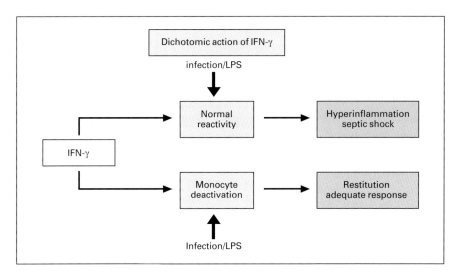

Fig. 6. Dichotomic action of IFN-γ in invasive infection and/or endotoxinemia.

the state of monocyte deactivation [15, 16]. Moreover, we demonstrated in a pilot trial that IFN-γ can also reverse 'immunoparalysis' in septic patients in vivo [3]. This was the first trial of a therapeutic intervention strategy in sepsis which used immune monitoring parameters as a guideline for treatment. Nine septic patients with 'immunoparalysis', defined as <30% HLA-DR + monocytes persisting for at least 2 days, were treated every day with 100 μg/d IFN-γ s.c. Treatment was continued until more than 50% of monocytes were HLA-DR + for 3 consecutive days (duration of treatment: mean 6 days, range 4–11 days). Although IFN-γ dramatically increases LPS-related mortality in the animal sepsis models, it was well tolerated by the patients with 'immunoparalysis' (fig. 6). Although the kinetics and the level of the response varied from individual to individual, monocyte recovery was observed in all 9 patients (fig. 7, 8). The slightly upregulated TNF plasma levels (23 pg/ml peak level on day 4 vs. 4 pg/ml before therapy) reflected the restoration of monocyte function in vivo and the recovery of an antimicrobial response (fig. 9). Eight of 9 patients recovered from sepsis shortly after treatment, although 2 of them relapsed at a later stage when IFN-γ treatment was discontinued. In parallel, the MOF score improved significantly. This pilot trial was not designed to prove effects on mortality but the results were promising.

Several groups have reported a diminished granulocyte function in immunoparalysis and monocytes from such patients had a diminished capacity to secrete G-CSF [Zuckermann et al., in prep.]. As G-CSF is not only important

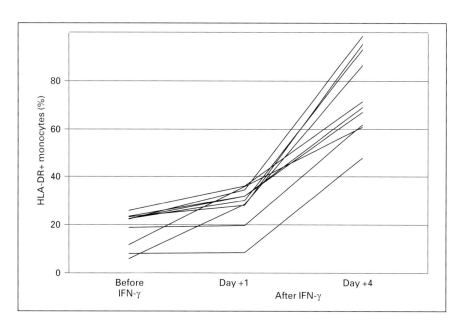

Fig. 7. IFN-γ therapy improved monocytic HLA-DR antigen expression in vivo in septic patients with 'immunoparalysis'. From Döcke et al. [3].

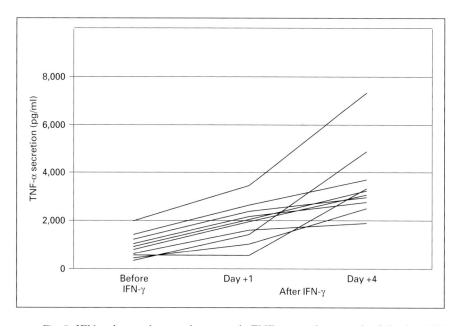

Fig. 8. IFN-γ therapy improved monocytic TNF-α secretion capacity following LPS stimulation ex vivo in septic patients with 'immunoparalysis'. From Döcke et al. [3].

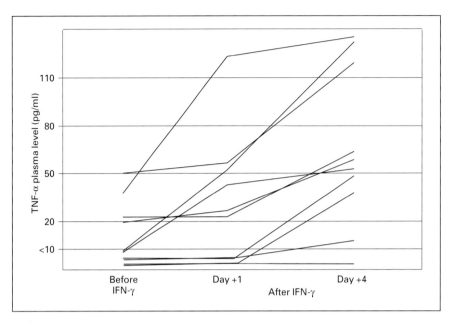

Fig. 9. The recovery of antimicrobial defense by IFN-γ therapy increased the TNF-α plasma levels in septic patients with 'immunoparalysis'.

for the granulocytopoiesis but also for the antimicrobial function and the survival time of mature granulocytes, we wondered whether G-CSF application would be a therapeutic alternative in sepsis (fig. 10). Twelve septic patients (abdominal/thoracic sepsis; three with ARDS) who had 'immunoparalysis' for at least 2 days were treated with G-CSF (3×300 μg) despite leukocytosis (mean 13 GPT/l). No significant side effects were observed. Leukocyte counts increased up to 75 GPT/l (mean 35 GPT/l) in 10/12 patients. Nine of them recovered from sepsis and this was associated with a decrease in procalcitonin plasma levels [Zuckermann et al., in prep.].

In a further approach we tested the feasibility of plasmapheresis in septic patients with 'immunoparalysis'. The rationale for this therapy is the removal of inhibitory factors from the plasma which cannot be achieved by dialysis. For this pilot trial we took 76 septic patients with 'immunoparalysis' for more than 4 days. 35 of them underwent plasmapheresis (3×4–5 liters). 18 of these 35 patients responded with stable recovery of monocyte function. 17 of the 18 responders survived but none of the 17 nonresponders did. Overall mortality was 52 vs. 80% in the untreated control group ($p < 0.01$) [9, 17].

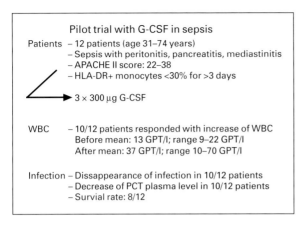

Pilot trial with G-CSF in sepsis

Patients   – 12 patients (age 31–74 years)
           – Sepsis with peritonitis, pancreatitis, mediastinitis
           – APACHE II score: 22–38
           – HLA-DR+ monocytes <30% for >3 days

                       3 × 300 µg G-CSF

WBC   – 10/12 patients responded with increase of WBC
             Before mean: 13 GPT/l; range 9–22 GPT/l
             After mean: 37 GPT/l; range 10–70 GPT/l

Infection – Dissappearance of infection in 10/12 patients
           – Decrease of PCT plasma level in 10/12 patients
           – Survial rate: 8/12

Fig. 10. Granulocyte colony-stimulating factor therapy in septic patients with leuko-cytosis.

Conclusions

What we have learned is that the immune system should be regarded as an organ which in ICU patients can be driven to failure just as a liver or a kidney can. The monitoring of other system functions (kidney, liver, blood pressure, lung, coagulation, etc.) is well established on ICUs, but monitoring of the immune system is still poorly developed. However, immune responsiveness is essential for the control of infections. In this sense the failure of immunotherapeutic approaches in sepsis introduced without immune monitoring is perhaps scarcely surprising. The data from our pilot trials suggest that immunostimulation in septic patients with defined 'immunoparalysis' may provide a novel approach. Controlled multicentre trials will be necessary to test this hypothesis. In the past poor standardisation of the immune monitoring made multicentre trials very difficult. However, this problem has been largely obviated by the recent development of rigorously standardised tests for immune parameters. On the one hand, immune competence can be measured by a standardised HLA-DR flow cytometric measurement and by a semiautomatic ex vivo whole blood TNF secretion assay, while on the other inflammation and tissue injury can be detected by measurement of plasma TNF and IL-6 levels. Moreover, it has been shown that measurement of procalcitonin plasma levels is helpful for early detection of invasive bacterial/fungal infections (fig. 11).

In our view, preemptive therapy of high-risk patients would be a much better approach than the treatment of established sepsis. Using measurement

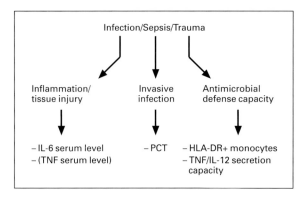

Fig. 11. Recommendations for a differential immune monitoring in ICU patients.

of immunocompetence, such as monocytic HLA-DR expression or whole endotoxin-induced blood TNF/IL-12 release, it is possible to detect patients with a high risk of developing infectious complications after major surgery and trauma as early as 2–3 days postoperatively. After an initial downregulation of the monocytic system which is seen in almost all patients after major surgery and trauma, immune function recovers in most patients. Persistence of the immune deactivation for several days is associated with an increasing risk to develop infectious complications. By day 2 of immunosuppression the predictive value is already >90%. It should therefore be possible to accurately select the high-risk patients who may benefit from a preemptive immunostimulatory therapy. With the availability of standardised immune monitoring procedures it should now be possible to design appropriate clinical trials.

References

1   Natanson C, Hoffman WD, Suffredini AF, Eichacker PQ, Danner RL: Selected treatment strategies for septic shock based on proposed mechanisms of pathogenesis. Ann Intern Med 1994;120: 771–783.
2   Zeni F, Freeman B, Natanson C: Anti-inflammatory therapies to treat sepsis and septic shock: A reassessment. Crit Care Med 1997;25:1095–1100.
3   Döcke WD, Randow F, Syrbe U, Krausch D, Asadullah K, Reinke P, Volk HD, Kox W: Monocyte deactivation in septic patients: Restoration by IFN-γ treatment. Nat Med 1997;3:678–681.
4   Echtenacher B, Falk W, Mannel DN, Krammer PH: Requirement of endogenous tumor necrosis factor/cachectin for recovery from experimental peritonitis. J Immunol 1990;145:3762–3766.
5   Pfeffer K: Mice deficient for the p55 kD tumor necrosis factor receptor are resistant to endotoxic shock, yet succumb to *L. monocytogenes* infection. Cell 1993;73:457–467.
6   Jack RS, Fan X, Bernheiden M, Rune G, Ehlers M, Weber A, Kirsch G, Mentel R, Fürll B, Freudenberg M, Schmitz G, Stelter F, Schütt C: Lipopolysaccharide binding protein is required to combat a murine gram-negative bacterial infection. Nature 1997;389:742–745.

7   Stroehlein MA, Fraunberger P, Stets P, Allgayer H, Tarabichi A, Delanoff C, Grützner KU, Faist E, Schildberg FW, Heiss MM: Evidence for a Th1 or Th2 respnse in autologous versus alogeneic blood transfusion associated immunomodulation; in Faist E (ed): The Immune Consequences of Trauma, Shock and Sepsis. Bologna, Monduzzi Editore, 1996, pp 287–292.

8   Volk HD, Reinke P, Falck P, Staffa G, Briedigkeit H, von Baehr R: Diagnostic value of an immune monitoring program for the clinical management of immunosuppressed patients with septic complications. Clin Transplant 1989;3:246–252.

9   Döcke WD, Reinke P, Syrbe U, Platzer C, Asadullah K, Krausch D, Zuckermann H, Volk HD: Immunoparalysis in sepsis – From phenomenon to treatment strategies. Tex Med 1997;9:55–65.

10   Randow F, Syrbe U, Meisel C, Krausch D, Zuckermann H, Platzer C, Volk HD: Mechanism of endotoxin desensitization: Involvement of IL-10 and TGF. J Exp Med 1995;5:1887–1892.

11   Woiciechowsky Ch, Asadullah K, Nestler D, Eberhardt B, Platzer C, Schöning B, Glöckner F, Lanksch WR, Volk HD, Döcke WD: Sympathetic activation triggers systemic IL-10 release in immunodepression induced by brain injury. Nat Med 1998;4:808–813.

12   Eskdale J, Gallagher G, Verweij CL, Keijsers V, Westendorp RG, Huizinga TW: Interleukin 10 secretion in relation to human IL-10 locus haplotypes. Proc Natl Acad Sci USA 1998;95:9465–9470.

13   van Dissel J, van Langevelde P, Westendorp RG, Kwappenberg K, Frolich M: Anti-inflammatory cytokine profile and mortality in febrile patients. Lancet 1998;351:950–953.

14   Asadulah K, Sterry W, Stephanek K, Leupold M, Jasulaitis D, Audring H, Volk HD, Sterry W, Döcke WD: IL-10 is a key cytokine in psoriasis. Proof of principle by IL-10 therapy: A new therapeutic approach. J Clin Invest 1998;101:1–12.

15   Randow F, Döcke WD, Bundschuh D, Hartung T, Wendel A, Volk HD: In vitro prevention and reversal of LPS desensitization by IFNg, IL-12, and GM-CSF. J Immunol 1997;158:2911–2918.

16   Bundschuh D, Barsig J, Hartung T, Randow F, Döcke WD, Volk HD, Wendel A: GM-CSF and interferon-gamma restore the systemic TNF alpha response to endotoxin in lipopolysaccaride-desensitized mice. J Immunol 1997;158:2862–2871.

17   Reinke P: Plasmapheresis in the therapy of septic disease. Int J Artif Organs 1996;19:127–128.

Hans-Dieter Volk, Institute of Medical Immunology, Charité-Campus Mitte,
Humboldt-University Berlin, D–10098 Berlin (Germany)

Author Index

Blunck, R. 5
Brandenburg, K. 5

Döcke, W.-D. 162
Dziarski, R. 83

Echtenacher, B.
  141

Gregory, C.D. 122
Gupta, D. 83

Jack, R.S. 1

Kitchens, R.L. 61

Latz, E. 42

Männel, D.N. 141

Reinke, P. 162

Schromm, A.B. 5

Schumann, R.R.
  42
Seydel, U. 5
Stelter, F. 25

Tapping, R.I. 108
Tobias, P.S. 108

Ulmer, A.J. 83

Volk, H.-D. 162

Subject Index

Apoptosis
  caspase signaling 123, 124
  CD14 role
    antibody inhibition studies 129
    cytokine production 134, 135
    ligands
      bridges 131
      intercellular adhesion molecule-3
        131, 132
      phosphatidylserine 130, 131
    signal transduction 132–135
    transfection studies 129
  cell surface characteristics
    charge 125
    ligands 126, 127
    phosphatidylserine externalization
      125, 126
  macrophage receptors, overview 127, 128
  morphological changes 123
  phagocytosis 124, 125, 134
  physiological functions 122, 123

Caspases, apoptosis roles 123, 124
CD14
  apoptosis role, *see* Apoptosis
  cellular distribution 25, 62
  coreceptors, *see also* specific receptors
    candidates 19
    mutation of putative interaction sites
      33, 34
    radiolabeling studies 114
  functional models 93, 94

gene structure 26
glycosylation 26, 29, 109
gram-positive bacteria component
  binding
    CD14-negative cell activation by soluble
      receptor complexes 90, 92, 93, 96, 97
    lipoarabinomannan binding 90
    lipotechoic acid binding, *see*
      Lipotechoic acid
    overview 2, 3, 27, 33, 53, 72, 73, 83
    peptidoglycan binding, *see*
      Peptidoglycan
    signal transduction pathways
      97, 98, 100
    transcription factors and gene
      activation 100
knockout mice 63, 64
lipopolysaccharide binding
  affinity 27
  binding site
    antibody inhibition studies 31, 32,
      111, 112
    membrane-bound vs soluble forms
      67, 68
    mutation analysis 29–31, 67, 111
    peptide-binding studies 32
    protection studies 31
  concentration dependence of effects 3
lipopolysaccharide-binding protein
  delivery of endotoxins
    aggregates 66
    monomers 66, 67

CD14, apoptosis role, affinity,
  lipopolysaccharide-binding protein
  delivery of endotoxins (continued)
    opsonized gram-negative bacteria
      64–66
    overview 7, 18–20, 26, 42, 43, 45,
      48–51, 53, 109, 110
    overview 1
    partial structure inhibition studies
      70, 71
    stoichiometry 26, 27, 66
  lipopolysaccharide internalization and
    signal transduction 68–70, 95
  membrane-bound form, *see* mCD14
  pattern recognition 53, 74–76, 93, 94
  phospholipid
    binding 74, 130, 131
    exchange 27, 51, 53, 113
  sequence analysis 26
  soluble form, *see* sCD14
  spirochete lipoprotein binding
    74
  tissue distribution 25, 26
CD36, apoptosis role 135
CD55, CD14 coreceptor 19
Complement receptor 3
  CD14 coreceptor 34, 96, 97, 133, 134
  tumor necrosis factor-α release
    mediation 149, 150
Complement receptor 4
  CD14 coreceptor 96, 133, 134
  tumor necrosis factor-α release mediation
    149, 150

Detergent insoluble glycolipid-enriched
  domain, mCD14 association 95

Endothelial cell, activation by sCD14
  complexes 92, 108, 109, 115
Endotoxin, *see* Lipid A, Lipopolysaccharide

Fgr, signal transduction in CD14
  activation 97, 98

Glycosylation, CD14 26, 29, 109
Granulocyte colony-stimulating factor,
  immunoparalysis therapy 172, 174

Hck, signal transduction in CD14
  activation 97, 98
High-density lipoprotein,
  lipopolysaccharide-binding protein
  association and lipopolysaccharide
  detoxification 50, 51

Immunoparalysis
  monocyte deactivation mechanisms
    apoptosis 166
    immunosuppression induction
      165, 171
    inflammatory cytokines 166, 168–170,
      175
    monocyte phenotypes 165
    stress 167, 168
  pathogens 165
  treatment
    granulocyte colony-stimulating factor
      therapy 172, 174
    immunosuppression modulation 171
    interferon-γ therapy 171, 172
    monocyte function monitoring
      171, 176
    plasmapheresis 174
Intercellular adhesion molecule-3, apoptosis
  role 127, 131, 132
Interferon-γ, immunoparalysis therapy
  171, 172
Interleukin-6, levels in immunoparalysis
  170
Interleukin-10
  immunoparalysis role 166, 168, 169
  psoriasis treatment 170

Lipid A
  analog binding to recognition proteins 71
  antagonists 10
  phosphorylation 17
  structure
    chemical structure 9, 10
    critical micellar concentration 11
    lamellar structures 11–13
    molecular modeling from crystal
      structure 14, 15
    phase states 13, 14
    signal transduction role 19, 20

toxicity effects 17
water content 15, 16
Lipoarabinomannan, CD14 binding 90
Lipopolysaccharide, *see also* Lipid A
  bacterial functions 1, 5
  CD14 binding, *see* CD14
  endothelial cell response 108
  internalization and signal transduction
    68–70, 95
  membrane interaction analysis 16
  partial structure inhibition studies 70, 71
  structure
    chemical structure 7–9
    micelle formation 2, 6
    molecular modeling from crystal
      structure 14, 15
    overview 5, 7
    phase states 13, 14, 17
    polymerization 6, 11–13
    signal transduction role 19, 20
    toxicity effects 9, 10, 17
    water content 15, 16
  toxicity
    aggregation effects 6, 7
    enhancement by lipopolysaccharide-
      binding protein 7
    mechanisms 5, 61, 62, 108
    species variations 5, 6, 9, 10
Lipopolysaccharide-binding protein
  binding site 44, 45
  catalytic mechanism of lipopolysaccharide
    transfer 49
  concentration
    acute phase upregulation 44, 45, 47
    dependence of effects 43, 45
    serum 44
  functions
    endotoxin micelle recognition 47, 48
    lipopolysaccharide delivery to CD14
      aggregates 66
      monomers 66, 67
      overview 7, 18–20, 26, 42, 43, 45,
        48–51, 53, 109, 110
    monomerization of lipopolysaccharide
      48, 53, 66, 67
    phospholipid transfer 51, 53, 113
  innate immune system role 42, 43

knockout mouse 4, 164
lipid specificity 18, 44, 53
lipopolysaccharide toxicity
  enhancement 7
lipoprotein association and
  lipopolysaccharide detoxification
    50, 51
sequence homology with other
  proteins 44, 45, 62
signaling role with CD14 18–20
structure 43, 44
Lipotechoic acid, CD14 binding
  stimulatory activity 74
  structural requirements 89
Lyn, signal transduction in CD14
  activation 97, 98

Major histocompatibility complex, CD14
  comparison 76
mCD14, *see also* CD14
  anchor role in signaling 68
  cellular distribution 25, 62
  detergent insoluble glycolipid-enriched
    domain association 95
  signal transduction 33–36, 94, 95
  transfection experiments conferring
    lipopolysaccharide response 62, 63, 92
Mitogen-activated protein kinase, signal
  transduction in CD14 activation
    97, 98, 100
Monocyte deactivation, *see*
  Immunoparalysis

Peptidoglycan
  CD14 binding
    antibody inhibition 86, 87
    binding sites 88, 89
    discovery 85, 86
    kinetics 87
    lipopolysaccharide-binding protein
      effects 87, 88
    muramyl dipeptide binding studies 73
    polymeric structure of ligand 88
    transcription factor activation 100
  function 85
  structure 85
Phagocytosis, apoptosis 124, 125, 134

Phosphatidylserine, externalization in
    apoptosis 125, 126
Plasmapheresis, immunoparalysis therapy
    174

sCD14, *see also* CD14
    activation of CD14-negative cells
        endothelial cells 108, 109, 115
        gram-positive bacterial ligands 90, 92,
            93, 96, 97
        mechanism 113–116
        myeloid cells 110, 111
        structure-function analysis of activation
            111–113
    concentration
        blood 28, 62, 90, 115
        dependence of effects 64, 111
    expression in disease 28, 29
    isoforms 28, 29
    neutralization of lipopolysaccharide
        27, 64, 111

Toll-like receptor 2, CD14 coreceptor 19, 34,
    35, 49, 72, 94, 96, 114, 115
Toll-like receptor 4
    CD14 coreceptor 35, 49, 72, 75, 94,
        96, 115
    functions 114
Transcription factors, activation by CD14
    interaction with gram-positive bacterial
    components 100, 101

Tumor necrosis factor-α
    antibody neutralization in sepsis 145, 146
    autoimmune disease role 152, 153
    biological activity 141–144
    cell types in production
        macrophages 147, 148
        mast cells 148, 149, 153
        monocytes 147
        neutrophils 148
    discovery 141
    early release
        functions in disease 151, 152
        stimuli 149, 150
    gene 142
    immunoparalysis role 166, 168, 170
    knockout mice studies 143
    monocyte deactivation, *see*
        Immunoparalysis
    receptors 143, 144, 152
    release on CD14 binding to
        lipopolysaccharide, overview 1, 146
    sepsis
        depression of expression in chronic
            sepsis 164
        protection effects 146, 147, 163
        toxicity 144, 145
    structure 142
    therapeutic targeting 142, 145, 153, 163

Vascular endothelial cells, activation by
    sCD14 complexes 92